TEST*		NONPARAMETRIC MEASURE OF CORRELATION (Chap. 9)
k-sample case		
Related samples (Chap. 7)	Independent samples (Chap. 8)	
chran Q test, pp. 161–166	χ^2 test for k independent samples, pp. 175–179	Contingency coefficient: C, pp. 196–202
dman two-way nalysis of ariance, pp. 66–172	Extension of the median test, pp. 179–184 Kruskal-Wallis one-way analysis of variance, pp. 184–193	Spearman rank correlation coefficient: r_S, pp. 202–213 Kendall rank correlation coefficient: τ, pp. 213–223 Kendall partial rank correlation coefficient: $\tau_{xy.z}$, pp. 223–229 Kendall coefficient of concordance: W, pp. 229–238

† The Wilcoxon test requires ordinal measurement not only within pairs, as is required for the sign test, but also of the differences between pairs. See the discussion on pp. 75–76.

311.2
Si1n

82931

Nonparametric Statistics

FOR THE BEHAVIORAL SCIENCES

John F. Dashiell was Consulting Editor of this series from its inception in 1931 until January 1, 1950. Clifford T. Morgan was Consulting Editor of this series from January 1, 1950 until January 1, 1959.

Nonparametric Statistics

FOR THE BEHAVIORAL SCIENCES

SIDNEY SIEGEL

Research Professor of Psychology
The Pennsylvania State University

McGRAW-HILL BOOK COMPANY

New York   Toronto   London

1956

NONPARAMETRIC STATISTICS: For the Behavioral Sciences

Library of Congress Catalog Card Number 56-8185

24 25 26 – MAMM – 7 5 4 3 2

ISBN 07-057348-4

To Jay

PREFACE

I believe that the nonparametric techniques of hypothesis testing are uniquely suited to the data of the behavioral sciences. The two alternative names which are frequently given to these tests suggest two reasons for their suitability. The tests are often called "distribution-free," one of their primary merits being that they do not assume that the scores under analysis were drawn from a population distributed in a certain way, e.g., from a normally distributed population. Alternatively, many of these tests are identified as "ranking tests," and this title suggests their other principal merit: nonparametric techniques may be used with scores which are not exact in any numerical sense, but which in effect are simply ranks. A third advantage of these techniques, of course, is their computational simplicity. Many believe that researchers and students in the behavioral sciences need to spend more time and reflection in the careful formulation of their research problems and in collecting precise and relevant data. Perhaps they will turn more attention to these pursuits if they are relieved of the necessity of computing statistics which are complicated and time-consuming. A final advantage of the nonparametric tests is their usefulness with small samples, a feature which should be helpful to the researcher collecting pilot study data and to the researcher whose samples must be small because of their very nature (e.g., samples of persons with a rare form of mental illness, or samples of cultures).

To date, no source is available which presents the nonparametric techniques in usable form and in terms which are familiar to the behavioral scientist. The techniques are presented in various mathematics and statistics publications. Most behavioral scientists do not have the mathematical sophistication required for consulting these sources. In addition, certain writers have presented summaries of these techniques in articles addressed to social scientists. Notables among these are Blum and Fattu (1954), Moses (1952a), Mosteller and Bush (1954), and Smith (1953). Moreover, some of the newer texts on statistics for social scientists have contained chapters on nonparametric methods. These include the texts by Edwards (1954), McNemar (1955), and Walker and Lev (1953). Valuable as these sources are, they have typically either been highly selective in the techniques presented or have not included the tables of significance values which are used in the application of the various tests. Therefore I have felt that a text on the nonparametric methods would be a desirable addition to the literature formed by the sources mentioned.

In this book I have presented the tests according to the research design for which each is suited. In discussing each test, I have attempted to indicate its "function," i.e., to indicate the sort of data to which it is applicable, to convey some notion of the rationale or proof underlying the test, to explain its computation, to give examples of its application in behavioral scientific research, and to compare the test to its parametric equivalent, if any, and to any nonparametric tests of similar function.

The reader may be surprised at the amount of space given to examples of the use of these tests, and even astonished at the repetitiousness which these examples introduce. I may justify this allocation of space by pointing out that (a) the examples help to teach the computation of the test, (b) the examples illustrate the application of the test to research problems in the behavioral sciences, and (c) the use of the same six steps in every hypothesis test demonstrates that identical logic underlies each of the many statistical techniques, a fact which is not well understood by many researchers.

Since I have tried to present all the raw data for each of the examples, I was not able to draw these from a catholic group of sources. Research publications typically do not present raw data, and therefore I was compelled to draw upon a rather parochial group of sources for most examples —those sources from which raw data were readily available. The reader will understand that this is an apology for the frequency with which I have presented in the examples my own research and that of my immediate colleagues. Sometimes I have not found appropriate data to illustrate the use of a test and therefore have "concocted" data for the purpose.

In writing this book, I have become acutely aware of the important influence which various teachers and colleagues have exercised upon my thinking. Professor Quinn McNemar gave me fundamental training in statistical inference and first introduced me to the importance of the assumptions underlying various statistical tests. Professor Lincoln Moses has enriched my understanding of statistics, and it was he who first interested me in the literature of nonparametric statistics. My study with Professor George Polya yielded exciting insights in probability theory. Professors Kenneth J. Arrow, Albert H. Bowker, Douglas H. Lawrence, and the late J. C. C. McKinsey have each contributed significantly to my understanding of statistics and experimental design. My comprehension of measurement theory was deepened by my research collaboration with Professors Donald Davidson and Patrick Suppes.

This book has benefited enormously from the stimulating and detailed suggestions and criticisms which Professors James B. Bartoo, Quinn McNemar, and Lincoln Moses gave me after each had read the manuscript. I am greatly indebted to each of them for their valuable gifts of time and knowledge. I am also grateful to Professors John F. Hall and Robert E. Stover, who encouraged my undertaking to write this book and who contributed helpful critical comments on some of the chapters. Of course, none of these persons is in any way responsible for the faults which remain; these are entirely my responsibility, and I should be grateful if any readers who detect errors and obscurities would call my attention to them.

Much of the usefulness of this book is due to the generosity of the many authors and publishers who have kindly permitted me to adapt or reproduce tables and other material originally presented by them. I have mentioned each source where the materials appear, and I also wish to mention here my gratitude to Donovan Auble, Irvin L. Child, Frieda Swed Cohn, Churchill Eisenhart, D. J. Finney, Milton Friedman, Leo A. Goodman, M. G. Kendall, William Kruskal, Joseph Lev, Henry B. Mann, Frank J. Massey, Jr., Edwin G. Olds, George W. Snedecor, Helen M. Walker, W. Allen Wallis, John E. Walsh, John W. M. Whiting, D. R. Whitney, and Frank Wilcoxon, and to the Institute of Mathematical Statistics, the American Statistical Association, *Biometrika*, the American Psychological Association, Iowa State College Press, Yale University Press, the Institute of Educational Research at Indiana University, the American Cyanamid Company, Charles Griffin & Co., Ltd., John Wiley & Sons, Inc., and Henry Holt and Company, Inc. I am indebted to Professor Sir Ronald A. Fisher, Cambridge, to Dr. Frank Yates, Rothamsted, and to Messrs. Oliver and Boyd, Ltd., Edinburgh, for permission to reprint Tables No. III and IV from their book *Statistical Tables for Biological, Agricultural, and Medical Research*.

My greatest personal indebtedness is to my wife, Dr. Alberta Engvall Siegel, without whose help this book could not have been written. She has worked closely with me in every phase of its planning and writing. I know it has benefited not only from her knowledge of the behavioral sciences but also from her careful editing, which has greatly enhanced any expository merits the book may have.

SIDNEY SIEGEL

CONTENTS

GLOSSARY OF SYMBOLS

A	Upper left-hand cell in a 2×2 table; number of cases observed in that cell.
α	Alpha. Level of significance = probability of a Type I error.
B	Upper right-hand cell in a 2×2 table; number of cases observed in that cell.
β	Beta. Power of the test = probability of a Type II error.
C	Lower left-hand cell in a 2×2 table; number of cases observed in that cell.
C	Contingency coefficient.
Chi square	A random variable which follows the chi-square distribution, certain values of which are shown in Table C of the Appendix.
χ^2	A statistic whose value is computed from observed data.
χ_r^2	The statistic in the Friedman two-way analysis of variance by ranks.
d_i	A difference score, used in the case of matched pairs, obtained for any pair by subtracting the score of one member from that of the other.
df	Degrees of freedom.
D	Lower right-hand cell in a 2×2 table; number of cases observed in that cell.
D	The maximum difference between the two cumulative distributions in the Kolmogorov-Smirnow test.
E_{ij}	Under H_0, the expected number of cases in the ith row and the jth column in a χ^2 test.
F	Frequency, i.e., number of cases.
F	The F test: the parametric analysis of variance.
$F_0(X)$	Under H_0, the proportion of cases in the population whose scores are equal to or less than X. This is a statistic in the Kolmogorov-Smirnov test.
g	In the Moses test, the amount by which an observed value of s_h exceeds $n_C - 2h$, where $n_C - 2h$ is the minimum span of the ranks of the control cases.
G_j	In the Cochran Q test, the total number of "successes" in the jth column (sample).
h	In the Moses test, the predetermined number of extreme control ranks which are dropped from each end of the span of control ranks before s_h is determined.
H	The statistic used in the Kruskal-Wallis one-way analysis of variance by ranks.
H_0	The null hypothesis.
H_1	The alternative hypothesis, the operational statement of the research hypothesis.
i	A variable subscript, usually denoting rows.

j	A variable subscript, usually denoting columns.
K	In the Kolmogorov-Smirnov test, the number of observations which are equal to or less than X.
K_D	In the Kolmogorov-Smirnov test, the numerator of D.
L_i	In the Cochran Q test, the total number of "successes" in the ith row.
μ	Mu. The population mean.
μ_0	The population mean under H_0.
μ_1	The population mean under H_1.
n	The number of independently drawn cases in a single sample.
N	The total number of independently drawn cases used in a statistical test.
O_{ij}	The observed number of cases in the ith row and the jth column in a χ^2 test.
p	Probability associated with the occurrence under H_0 of a value as extreme as or more extreme than the observed value.
P	In the binomial test, the proportion of "successes."
Q	In the binomial test, $1 - P$.
Q	The statistic used in the Cochran test.
r	The number of runs.
r	The Pearson product-moment correlation coefficient.
r	The number of rows in a $k \times r$ table.
R_j	The sum of the ranks in the jth column or sample.
r_S	The Spearman rank correlation coefficient.
$r_{S_{av}}$	The mean of several r_S's.
s	In the Kendall W, the sum of the squares of the deviations of the R_j from the mean value of R_j.
s'	In the Moses test, the span or range of the ranks of the control cases.
s_h	In the Moses test, the span or range of the ranks of the control cases after h cases have been dropped from each extreme of that range.
S	A statistic in the Kendall τ.
$S_N(X)$	In the Kolmogorov-Smirnov test, the observed cumulative step function of a random sample of N observations.
σ	Sigma. The standard deviation of the population. When a subscript is given, the standard error of a sampling distribution, for example, σ_U = the standard error of the sampling distribution of U.
σ^2	The variance of the population.
Σ	Summation of.
t	Student's t test, a parametric test.
t	The number of observations in any tied group.
T	In the Wilcoxon test, the smaller of the sums of like-signed ranks.
T	A correction factor for ties.
τ	Tau. The Kendall rank correlation coefficient.
$\tau_{xy.z}$	The Kendall partial rank correlation coefficient.
U	The statistic in the Mann-Whitney test.
U'	$U = n_1 n_2 - U'$, a transformation in the Mann-Whitney test.
W	The Kendall coefficient of concordance.
x	In the binomial test, the number of cases in one of the groups.
X	Any observed score.
$\bar{X}$	The mean of a sample of observations.
z	Deviation of the observed value from μ_0 when $\sigma = 1$. z is normally distributed. Probabilities associated under H_0 with the occurrence of values as extreme as various z's are given in Table A of the Appendix.

$\binom{a}{b}$ The binomial coefficient. $\binom{a}{b} = \dfrac{a!}{b!\,(a-b)!}$.   Table T of the Appendix gives binomial coefficients for N from 1 to 20.

!   Factorial. $N! = N(N-1)(N-2)\cdots 1$.   For example,

$$5! = (5)(4)(3)(2)(1) = 120$$

$0! = 1$.   Tables S of the Appendix gives factorials for N from 1 to 20.

$|X - Y|$ The absolute value of the difference between X and Y.   That is, the numerical value of the difference regardless of sign.   For example, $|5 - 3| = |3 - 5| = 2$.

$X > Y$ X is greater than Y.

$X < Y$ X is less than Y.

$X = Y$ X is equal to Y.

$X \geq Y$ X is equal to or greater than Y.

$X \leq Y$ X is equal to or less than Y.

$X \neq Y$ X is not equal to Y.

CHAPTER 1

INTRODUCTION

The student of the behavioral sciences soon grows accustomed to using familiar words in initially unfamiliar ways. Early in his study he learns that when a behavioral scientist speaks of *society* he is not referring to that leisured group of persons whose names appear in the society pages of our newspapers. He knows that the scientific denotation of the term *personality* has little or nothing in common with the teen-ager's meaning. Although a high school student may contemptuously dismiss one of his peers for having "no personality," the behavioral scientist can scarcely conceive such a condition. The student has learned that *culture*, when used technically, encompasses far more than aesthetic refinement. And he would not now be caught in the blunder of saying that a salesman "*uses*" *psychology* in persuading a customer to purchase his wares.

Similarly, the student has discovered that the field of *statistics* is quite different from the common conception of it. In the newspapers and in other journals of popular thinking, the statistician is represented as one who collects large amounts of quantitative information and then abstracts certain representative numbers from that information. We are all familiar with the notion that the determination of the average hourly wage in an industry or of the average number of children in urban American families is the statistician's job. But the student who has taken even one introductory course in statistics knows that description is only one function of statistics.

A central topic of modern statistics is *statistical inference*. Statistical inference is concerned with two types of problems: estimation of population parameters and tests of hypotheses. It is with the latter type, tests of hypotheses, that we shall be primarily concerned in this book.

Webster tells us that the verb "to infer" means "to derive as a consequence, conclusion, or probability." When we see a woman who wears no ring on the third finger of her left hand, we may *infer* that she is unmarried.

In statistical inference, we are concerned with how to draw conclusions about a large number of events on the basis of observations of a portion of them. Statistics provides tools which formalize and standardize our

1

procedures for drawing conclusions. For example, we might wish to determine which of three varieties of tomato sauce is most popular with American homemakers. Informally, we might gather information on this question by stationing ourselves near the tomato sauce counter at a grocery store and counting how many cans of each variety are purchased in the course of a day. Almost certainly the numbers of purchases of the three varieties will be unequal. But can we *infer* that the one most frequently chosen on that day in that store by that day's customers is really the most popular among American homemakers? Whether we can make such an inference must depend on the margin of popularity held by the most frequently chosen brand, on the representativeness of the grocery store, and also on the representativeness of the group of purchasers whom we observed.

The procedures of statistical inference introduce order into any attempt to draw conclusions from the evidence provided by samples. The logic of the procedures dictates some of the conditions under which the evidence must be collected, and statistical tests determine how large the observed differences must be before we can have confidence that they represent real differences in the larger group from which only a few events were sampled.

A common problem for statistical inference is to determine, in terms of probability, whether observed differences between two samples signify that the populations sampled are themselves really different. Now whenever we collect two groups of scores by random methods we are likely to find that the scores differ to some extent. Differences occur simply because of the operations of chance. Then how can we determine in any given case whether the observed differences are merely due to chance or not? The procedures of statistical inference enable us to determine, in terms of probability, whether the observed difference is within the range which could easily occur by chance or whether it is so large that it signifies that the two samples are probably from two different populations. Another common problem is to determine whether it is likely that a sample of scores is from some specified population. Still another is to decide whether we may legitimately infer that several groups differ among themselves. We shall be concerned with each of these tasks for statistical inference in this book.

In the development of modern statistical methods, the first techniques of inference which appeared were those which made a good many assumptions about the nature of the population from which the scores were drawn. Since population values are "parameters," these statistical techniques are called *parametric*. For example, a technique of inference may be based on the assumption that the scores were drawn from a normally distributed population. Or the technique of inference may be based on the assumption that both sets of scores were drawn from populations having the same variance (σ^2) or spread of scores. Such techniques produce conclusions which contain qualifiers, e.g., "If the assumptions regarding the shape of the population(s) are valid, then we may conclude that . . . ."

More recently we have seen the development of a large number of techniques of inference which do not make numerous or stringent assumptions about parameters. These newer "distribution-free" or *nonparametric* techniques result in conclusions which require fewer qualifications. Having used one of them, we may say that "Regardless of the shape of the population(s), we may conclude that . . . ." It is with these newer techniques that we shall be concerned in this book.

Some nonparametric techniques are often called "ranking tests" or "order tests," and these titles suggest another way in which they differ from the parametric tests. In the computation of parametric tests, we add, divide, and multiply the scores from the samples. When these arithmetic processes are used on scores which are not truly numerical, they naturally introduce distortions in those data and thus throw in doubt any conclusions from the test. Thus it is permissible to use the parametric techniques only with scores which are truly numerical. Many nonparametric tests, on the other hand, focus on the order or ranking of the scores, not on their "numerical" values, and other nonparametric techniques are useful with data for which even ordering is impossible (i.e., with classificatory data). Whereas a parametric test may focus on the difference between the means of two sets of scores, the equivalent nonparametric test may focus on the difference between the medians. The computation of the mean requires arithmetic manipulation (addition and then division); the computation of the median requires only counting. The advantages of order statistics for data in the behavioral sciences (in which "numerical" scores may be precisely numerical in appearance only!) should be apparent. We shall discuss this point at greater length in Chap. 3, in which the parametric and nonparametric tests are contrasted.

Of the nine chapters contained in this book, six are devoted to the presentation of the various nonparametric statistical tests. The tests are assigned to chapters according to the research design for which they are appropriate. One chapter contains those tests which may be used when one wishes to determine whether a single sample is from a specified sort of population. Two chapters contain those tests which may be used when one wishes to compare the scores yielded by two samples; one of these chapters considers tests for two related samples, while the other considers tests for two independent samples. Similarly, two chapters are devoted to significance tests for k (3 or more) samples; one of these presents tests for k related samples and the other presents tests for k independent samples. The final chapter gives nonparametric measures of association, and the tests of significance which are useful with some of these.

Before the reader comes to these chapters, however, he will be confronted with two others in addition to the present one. The first of these, Chap. 2, is devoted to a general discussion of tests of hypotheses. Because this discussion is really little more than a summary statement about the elementary aspects of hypothesis testing, and because much of its vocabulary may be unfamiliar, some readers—especially those with little or no acquaintance with the theory of statistical inference—may find Chap. 2 a difficult one. We suggest that such persons would do well to turn to the references cited there for a more comprehensive treatment of the notions discussed. We hope, however, that for most readers Chap. 2 will provide sufficient background for understanding the balance of the book. The notions and vocabulary introduced in Chap. 2 are employed frequently and even repetitiously throughout the book, and therefore should become more familiar and meaningful as the reader repeatedly encounters them in succeeding chapters.

Chapter 3 discusses the choice of that statistical technique which is best suited for analyzing a given batch of data. This discussion includes a comparison of parametric and nonparametric statistical tests, and introduces the reader to the theory of measurement. Again the reader may find that he is facing much new material in a few pages. And again we suggest that the new material will become increasingly familiar as he progresses through the succeeding chapters.

We have tried to make this book fully comprehensible to the reader who has had only introductory work in statistics. It is presumed that the reader will have a speaking acquaintance with descriptive statistics (means, medians, standard deviations, etc.), with parametric correlational methods (particularly the Pearson product-moment correlation r), and with the basic notions of statistical inference and their use in the t test and in the analysis of variance. The reader who has had even limited experience with these statistics and statistical tests should find the references to them comprehensible.

Moreover, we have tried to make the book completely intelligible to the reader whose mathematical training is limited to elementary algebra. This orientation has precluded our presenting many derivations. Where possible we have tried to convey an "intuitive" understanding of the rationale underlying a test, and have thought that this understanding will be more useful than an attempt to follow the derivation. The more mathematically sophisticated reader will want to pursue the topics of this book by turning to the sources to which we have made reference.

Readers whose mathematical training is limited, and especially readers whose educational experience has been such that they have developed negative emotional responses to symbols, often find statistics books difficult because of the extensive use of symbols. Such readers may find that much of this difficulty will disappear if they read more slowly than is their custom. It is not to be expected that a reader schooled in the behavioral sciences can maintain the same fast clip in reading a statistics book that he maintains in reading a book on, say, personality or on intergroup hostility or on the role of geography in cultural differences. Statistical writing is more condensed than most social scientific writing—we use symbols for brevity as well as for exactness—and therefore it requires slower reading. The reader who finds symbols difficult may also be aided by the glossary which is included. That glossary summarizes the meanings of the various symbols used in the book.

The reader with limited mathematical training may also find the examples especially helpful: an example is given of the use in research of every statistical test. One reason that the extensive use of symbols makes material more difficult may be that symbols are general or abstract terms, which acquire a variety of specific meanings in a variety of specific cases. Thus when we speak of k samples we mean any number of samples, 3 or 4 or 8 or 5 or any other number. In the examples, of course, the symbols each acquire a specific numerical value, and thus the examples may serve to "concretize" the discussion for the reader.

The examples also serve to illustrate the role and importance of statistics in the research of the behavioral scientist. This may be their most useful function, for we have addressed this book to the researcher whose primary interest is in the substance or topical fields of the social sciences rather than in their methodology. The examples demonstrate the intimate interrelation of substance and method in the behavioral sciences.

CHAPTER 2

THE USE OF STATISTICAL TESTS IN RESEARCH

In the behavioral sciences we conduct research in order to determine the acceptability of hypotheses which we derive from our theories of behavior. Having selected a certain hypothesis which seems important in a certain theory, we collect empirical data which should yield direct information on the acceptability of that hypothesis. Our decision about the meaning of the data may lead us to retain, revise, or reject the hypothesis and the theory which was its source.

In order to reach an objective decision as to whether a particular hypothesis is confirmed by a set of data, we must have an objective procedure for either rejecting or accepting that hypothesis. Objectivity is emphasized because one of the requirements of the scientific method is that one should arrive at scientific conclusions by methods which are public and which may be repeated by other competent investigators.

This objective procedure should be based on the information we obtain in our research, and on the risk we are willing to take that our decision with respect to the hypothesis may be incorrect.

The procedure usually followed involves several steps. Here we list these steps in their order of performance; this and the following chapter are devoted to discussing each in some detail.

i. State the null hypothesis (H_0).

ii. Choose a statistical test (with its associated statistical model) for testing H_0. From among the several tests which might be used with a given research design, choose that test whose model most closely approximates the conditions of the research (in terms of the assumptions which qualify the use of the test) and whose measurement requirement is met by the measures used in the research.

iii. Specify a significance level (α) and a sample size (N).

iv. Find (or assume) the sampling distribution of the statistical test under H_0.

v. On the basis of (ii), (iii), and (iv) above, define the region of rejection.

vi. Compute the value of the statistical test, using the data obtained from the sample(s). If that value is in the region of rejection, the decision is to reject H_0; if that value is outside the region of rejection, the decision is that H_0 cannot be rejected at the chosen level of significance.

A number of statistical tests are presented in this book. In most presentations, one or more examples of the use of the test in research are given. Each example follows the six steps given above. An understanding of the reason for each of these steps is central to an understanding of the role of statistics in testing a research hypothesis.

i. THE NULL HYPOTHESIS

The first step in the decision-making procedure is to state the null hypothesis (H_0). The *null hypothesis* is a hypothesis of no differences. It is usually formulated for the express purpose of being rejected. If it is rejected, the alternative hypothesis (H_1) may be accepted. The *alternative hypothesis* is the operational statement of the experimenter's research hypothesis. The *research hypothesis* is the prediction derived from the theory under test.

When we want to make a decision about differences, we test H_0 against H_1. H_1 constitutes the assertion that is accepted if H_0 is rejected.

Suppose a certain social scientific theory would lead us to predict that two specified groups of people differ in the amount of time they spend in reading newspapers. This prediction would be our research hypothesis. Confirmation of that prediction would lend support to the social scientific theory from which it was derived. To test this research hypothesis, we state it in operational form as the alternative hypothesis, H_1. H_1 would be that $\mu_1 \neq \mu_2$, that is, that the mean amount of time spent in reading newspapers by the members of the two populations is unequal. H_0 would be that $\mu_1 = \mu_2$, that is, that the mean amount of time spent in reading newspapers by the members of the two populations is the same. If the data permit us to reject H_0, then H_1 can be accepted, and this would support the research hypothesis and its underlying theory.

The nature of the research hypothesis determines how H_1 should be stated. If the research hypothesis simply states that two groups will differ with respect to means, then H_1 is that $\mu_1 \neq \mu_2$. But if the theory predicts the *direction* of the difference, i.e., that one specified group will have a larger mean than the other, then H_1 may be either that $\mu_1 > \mu_2$ or that $\mu_1 < \mu_2$ (where $>$ means "greater than" and $<$ means "less than").

ii. THE CHOICE OF THE STATISTICAL TEST

The field of statistics has developed to the extent that we now have, for almost all research designs, alternative statistical tests which might be used in order to come to a decision about a hypothesis. Having alternative tests, we need some rational basis for choosing among them. Since this book is concerned with nonparametric statistics, the choice among (parametric and nonparametric) statistics is one of its central topics. Therefore the discussion of this point is reserved for a separate chapter; Chap. 3 gives an extended discussion of the bases for choosing among the various tests applicable to a given research design.

iii. THE LEVEL OF SIGNIFICANCE AND THE SAMPLE SIZE

When the null hypothesis and the alternative hypothesis have been stated, and when the statistical test appropriate to the research has been selected, the next step is to specify a level of significance (α) and to select a sample size (N).

In brief, this is our decision-making procedure: In advance of the data collection, we specify the set of all possible samples that could occur when H_0 is true. From these, we specify a subset of possible samples which are so extreme that the probability is very small, if H_0 is true, that the sample we actually observe will be among them. If in our research we then observe a sample which was included in that subset, we reject H_0.

Stated differently, our procedure is to reject H_0 in favor of H_1 if a statistical test yields a value whose associated probability of occurrence under H_0 is equal to or less than some small probability symbolized as α. That small probability is called *the level of significance*. Common values of α are .05 and .01. To repeat: if the probability associated with the occurrence under H_0, i.e., when the null hypothesis is true, of the particular value yielded by a statistical test is equal to or less than α, we reject H_0 and accept H_1, the operational statement of the research hypothesis.[1]

[1] In contemporary statistical decision theory, the procedure of adhering rigidly to an arbitrary level of significance, say .05 or .01, has been rejected in favor of the procedure of making decisions in terms of loss functions, utilizing such principles as the minimax principle (the principle of minimizing the maximum loss). For a discussion of this approach, the reader may turn to Blackwell and Girshick (1954), Savage (1954), or Wald (1950). Although the desirability of such a technique for arriving at decisions is clear, its practicality in most research in the behavioral sciences at present is dubious, because we lack the information which would be basic to the use of loss functions.

A common practice, which reflects the notion that different investigators and readers may hold different views as to the "losses" or "gains" involved in implementing a social scientific finding, is for the researcher simply to report the probability level associated with his finding, indicating that the null hypothesis may be rejected at that level.

From the discussion of significance levels which is given in this book, the reader should not infer that the writer believes in a rigid or hard-and-fast approach to the setting of significance levels. Rather, it is for heuristic reasons that significance levels are emphasized; such an exposition seems the best method of clarifying the role which the information contained in the sampling distribution plays in the decision-making procedure.

It can be seen, then, that α gives the probability of mistakenly or falsely rejecting H_0. This interpretation of α will be amplified when the Type I error is discussed.

Since the value of α enters into the determination of whether H_0 is or is not rejected, the requirement of objectivity demands that α be set in advance of the collection of the data. The level at which the researcher chooses to set α should be determined by his estimate of the importance or possible practical significance of his findings. In a study of the possible therapeutic effects of brain surgery, for example, the researcher may well choose to set a rather stringent level of significance, for the dangers of rejecting the null hypothesis improperly (and therefore unjustifiably advocating or recommending a drastic clinical technique) are great indeed. In reporting his findings, the researcher should indicate the actual probability level associated with his findings, so that the reader may use his own judgment in deciding whether or not the null hypothesis should be rejected. A researcher may decide to work at the .05 level, but a reader may refuse to accept any finding not significant at the .01, .005, or .001 levels, while another reader may be interested in any finding which reaches, say, the .08 or .10 levels. The researcher should give his readers the information they require by reporting, if possible, the probability level actually associated with the finding.

There are two types of errors which may be made in arriving at a decision about H_0. The first, the *Type I error*, is to reject H_0 when in fact it is true. The second, the *Type II error*, is to accept H_0 when in fact it is false.

The probability of committing a Type I error is given by α. The larger is α, the more likely it is that H_0 will be rejected falsely, i.e., the more likely it is that the Type I error will be committed. The Type II error is usually represented by β. α and β will be used here to indicate both the type of error and the probability of making that error. That is,

$$p \text{ (Type I error)} = \alpha$$
$$p \text{ (Type II error)} = \beta$$

Ideally, the specific values of both α and β would be specified by the experimenter before he began his research. These values would determine the size of the sample (N) he would have to draw for computing the statistical test he had chosen.

In practice, however, it is usual for α and N to be specified in advance. Once α and N have been specified, β is determined. Inasmuch as there is an inverse relation between the likelihood of making the two types of errors, a decrease in α will increase β for any given N. If we wish to reduce the possibility of both types of errors, we must increase N.

It should be clear that in any statistical inference a danger exists of committing one of the two alternative types of errors, and that the experimenter should reach some compromise which optimizes the balance between the probabilities of making the two errors. The various statistical tests offer the possibility of different balances. It is in achieving this balance that the notion of the power function of a statistical test is relevant.

The *power of a test* is defined as the probability of rejecting H_0 when it is in fact *false*. That is,

$$\text{Power} = 1 - \text{probability of Type II error} = 1 - \beta$$

The curves in Fig. 1 show that the probability of committing a Type II error (β) decreases as the sample size (N) increases, and thus that power increases with the size of N. Figure 1 illustrates the increase in power of the two-tailed test of the mean which comes with increasing sample sizes: $N = 4, 10, 20, 50,$ and 100. These samples are taken from normal populations with variance σ^2. The mean under the null hypothesis is symbolized here as μ_0.

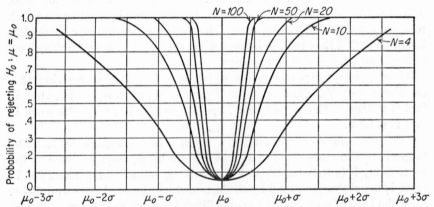

FIG. 1. Power curves of the two-tailed test at $\alpha = .05$ with varying sample sizes.

Figure 1 also shows that when H_0 is true, i.e., when the true mean $= \mu_0$, the probability of rejecting $H_0 = .05$. This is as it should be, inasmuch as $\alpha = .05$, and α gives the probability of rejecting H_0 when it is in fact true.

From this discussion it is important that the reader understand the following five points, which summarize what we have said about the selection of the level of significance and of the sample size:

1. The significance level α is the probability that a statistical test will yield a value under which the null hypothesis will be rejected when in fact it is true. That is, the significance level indicates the probability of committing the Type I error.

2. β is the probability that a statistical test will yield a value under

which the null hypothesis will be accepted when in fact it is false. That is, β gives the probability of committing the Type II error.

3. The power of a test, $1 - \beta$, tells the probability of rejecting the null hypothesis when it is false (and thus should be rejected).

4. Power is related to the nature of the statistical test chosen.[1]

5. Generally the power of a statistical test increases with an increase in N.

iv. THE SAMPLING DISTRIBUTION

When an experimenter has chosen a certain statistical test to use with his data, he must next determine what is the sampling distribution of the test statistic.

The sampling distribution is a theoretical distribution. It is that distribution we would get if we took *all possible* samples of the same size from the same population, drawing each randomly. Another way of saying this is to say that the sampling distribution is the distribution, *under H_0*, of all possible values that some statistic (say the sample mean, $\bar{X}$) can take when that statistic is computed from randomly drawn samples of equal size.

The sampling distribution of a statistic shows the probabilities under H_0 associated with various possible numerical values of the statistic. The probability "associated with" the occurrence of a particular value of the statistic under H_0 is *not* the exact probability of just that value. Rather, "the probability associated with the occurrence under H_0" is here used to refer to the probability of a particular value *plus* the probabilities of all more extreme possible values. That is, the "associated probability" or "the probability associated with the occurrence under H_0" is the probability of the occurrence under H_0 of a value *as extreme as or more extreme than* the particular value of the test statistic. In this book we shall have frequent occasion to use the above phrases, and in each case they shall carry the meaning given above.

Suppose we were interested in the probability that three heads would land up when three "fair" coins were tossed simultaneously. The sampling distribution of the number of heads could be drawn from the list of all possible results of tossing three coins, which is given in Table 2.1. The total number of possible events (possible combinations of H's and T's—heads and tails) is eight, only one of which is the event in which we are interested: the simultaneous occurrence of three H's. Thus the probability of the occurrence under H_0 of three heads on the toss of three coins is $\frac{1}{8}$. Here H_0 is the assertion that the coins are "fair," which means that

---

[1] Power is also related to the nature of H_1. If H_1 has direction, a one-tailed test is used. A one-tailed test is more powerful than a two-tailed test. This should be clear from the definition of power.

TABLE 2.1. POSSIBLE OUTCOMES OF THE TOSS OF THREE COINS

	Outcomes							
	1	2	3	4	5	6	7	8
Coin 1	H	H	H	H	T	T	T	T
Coin 2	H	H	T	T	H	H	T	T
Coin 3	H	T	H	T	H	T	H	T

for each coin the probability of a head occurring is equal to the probability of a tail occurring.

Thus the sampling distribution of all possible events has shown us the probability of the occurrence under H_0 of the event with which we are concerned.

It is obvious that it would be essentially impossible for us to use this method of imagining all possible results in order to write down the sampling distributions for even moderately large samples from large populations. This being the case, we rely on the authority of statements of "proved" mathematical theorems. These theorems invariably involve assumptions, and in applying the theorems we must keep the assumptions in mind. Usually these assumptions concern the distribution of the population and/or the size of the sample. An example of such a theorem is the *central-limit theorem*.

When a variable is normally distributed, its distribution is completely characterized by the mean and the standard deviation. This being the case, we know, for example, that the probability that an observed value of such a variable will differ from the mean by more than 1.96 standard deviations is less than .05. (The probabilities associated with any difference in standard deviations from the mean of a normally distributed variable are given in Table A of the Appendix.)

Suppose then we want to know, before the sample is drawn, the probability associated with the occurrence of a particular value of $\bar{X}$ (the arithmetic mean of the sample), i.e., the probability under H_0 of the occurrence of a value at least as large as a particular value of $\bar{X}$, when the sample is randomly drawn from some population whose mean μ and standard deviation σ we know. One version of the central-limit theorem states that:

If a variable is distributed with mean $= \mu$ and standard deviation $= \sigma$, and if random samples of size N are drawn, then the means of these samples, the $\bar{X}$'s, will be approximately normally distributed with mean μ and standard deviation $\sigma/\sqrt{N}$ for N sufficiently large.

In other words, if N is sufficiently large, we know that the sampling distribution of $\bar{X}$ (a) is approximately normal, (b) has a mean equal to the population mean μ, and (c) has a standard deviation which is equal to the population standard deviation divided by the square root of the sample size, that is, $\sigma_{\bar{x}} = \sigma/\sqrt{N}$.

For example, suppose we know that in the population of American college students, some psychological attribute, as measured by some test, is distributed with $\mu = 100$ and $\sigma = 16$. Now we want to know the probability of drawing a random sample of 64 cases from this population and finding that the mean score in that sample, $\bar{X}$, is as large as 104. The central-limit theorem tells us that the sampling distribution of $\bar{X}$'s of all possible samples of size 64 will be approximately normally distributed and will have a mean equal to 100 ($\mu = 100$) and a standard deviation equal to $\sigma/\sqrt{N} = 16/\sqrt{64} = 2$. We can see that 104 differs from 100 by two standard errors.[1] Reference to Table A reveals that the probability associated with the occurrence under H_0 of a value as large as such an observed value of $\bar{X}$, that is, of an $\bar{X}$ which is at least two standard errors above the mean ($z \geq 2.0$), is $p < .023$.

It should be clear from this discussion and this example that by knowing the sampling distribution of some statistic we are able to make probability statements about the occurrence of certain numerical values of that statistic. The following sections will show how we use such a probability statement in making a decision about H_0.

v. THE REGION OF REJECTION

The region of rejection is a region of the sampling distribution. The sampling distribution includes *all* possible values a test statistic can take under H_0; the region of rejection consists of a subset of these possible values, and is defined so that the probability under H_0 of the occurrence of a test statistic having a value which is in that subset is α. In other words, the region of rejection consists of a set of possible values which are so extreme that when H_0 is true the probability is very small (i.e., the probability is α) that the sample we actually observe will yield a value which is among them. The probability *associated* with any value in the region of rejection is equal to or less than α.

The location of the region of rejection is affected by the nature of H_1. If H_1 indicates the predicted direction of the difference, then a one-tailed test is called for. If H_1 does not indicate the direction of the predicted difference, then a two-tailed test is called for. One-tailed and two-tailed tests differ in the location (but not in the size) of the region of rejection. That is, in a one-tailed test the region of rejection is entirely at one end

[1] The standard deviation of a sampling distribution is usually called a *standard error*.

(or tail) of the sampling distribution. In a two-tailed test, the region of rejection is located at both ends of the sampling distribution.

The size of the region of rejection is expressed by α, the level of significance. If $\alpha = .05$, then the size of the region of rejection is 5 per cent of the entire space included under the curve in the sampling distribution.

One-tailed and two-tailed regions of rejection for $\alpha = .05$ are illustrated in Fig. 2. Observe that these two regions differ in location but not in total size.

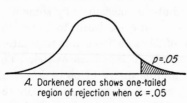

A. Darkened area shows one-tailed region of rejection when $\alpha = .05$

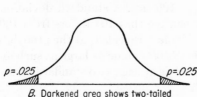

B. Darkened area shows two-tailed region of rejection when $\alpha = .05$

FIG. 2. Regions of rejection for one-tailed and two-tailed tests.

vi. THE DECISION

If the statistical test yields a value which is in the region of rejection, we reject H_0.

The reasoning behind this decision process is very simple. If the probability associated with the occurrence under the null hypothesis of a particular value in the sampling distribution is very small, we may explain the actual occurrence of that value in two ways: first, we may explain it by deciding that the null hypothesis is false, or second, we may explain it by deciding that a rare and unlikely event has occurred. In the decision process, we choose the first of these explanations. Occasionally, of course, the second may be the correct one. In fact, the probability that the second explanation is the correct one is given by α, for rejecting H_0 when in fact it is true is the Type I error.

When the probability associated with an observed value of a statistical test is equal to or less than the previously determined value of α, we conclude that H_0 is false. Such an observed value is called "significant." H_0, the hypothesis under test, is rejected whenever a "significant" result occurs. A "significant" value is one whose associated probability of occurrence under H_0 (as shown by the sampling distribution) is equal to or less than α.

EXAMPLE

In the discussions of the various nonparametric statistical tests, many examples of statistical decisions will be given in this book. Here we shall give just one example of how a statistical decision is reached, to illustrate the points made in this chapter.

EXAMPLE 15

Suppose we suspect a particular coin of being biased. Our suspicion is that the coin is biased to land with head up. To test this suspicion (which we here may dignify by calling it a "research hypothesis"), we decide to toss the coin 12 times and to observe the frequency with which head occurs.

i. *Null Hypothesis.* H_0: $p(H) = p(T) = \frac{1}{2}$. That is, for this coin there is no difference between the probability of the occurrence of a head, that is, $p(H)$, and the probability of the occurrence of a tail, that is, $p(T)$; the coin is "fair." H_1: $p(H) > p(T)$.

ii. *Statistical Test.* The statistical test which is appropriate to test this hypothesis is the binomial test, which is based on the binomial expansion. (This test is presented fully in Chap. 4.)

iii. *Significance Level.* In advance we decide to use $\alpha = .01$ as our level of significance. $N = 12 =$ the number of independent tosses.

iv. *Sampling Distribution.* The sampling distribution which gives the probability of obtaining x heads and $N - x$ tails under the null hypothesis (the hypothesis that the coin is in fact fair) is given by the binomial distribution function: $\dfrac{N!}{x!\,(N - x)!} P^x Q^{N-x}$; $x = 0, 1, 2, \ldots, N$. Table 2.2 shows the sampling distribution of x, the

TABLE 2.2. SAMPLING DISTRIBUTION OF x (NUMBER OF HEADS) FOR 2^{12} SAMPLES OF SIZE $N = 12$

Number of heads	Sampling distribution (Expected frequency of occurrence if 2^{12} samples of 12 tosses were taken)
12	1
11	12
10	66
9	220
8	495
7	792
6	924
5	792
4	495
3	220
2	66
1	12
0	1
	Total $= 2^{12} = 4,096$

number of heads. This sampling distribution shows that the most likely outcome of tossing a coin 12 times is to obtain 6 heads and 6 tails. Obtaining 7 heads and 5 tails is somewhat less likely but still quite probable. But the occurrence of 12 heads on 12 tosses is very unlikely indeed. The occurrence of 0 heads (12 tails) is equally unlikely.

v. *Rejection Region.* Since H_1 has direction, a one-tailed test will be used, and thus the region of rejection is entirely at one end of the sampling distribution. The region of rejection consists of all values of x (number of heads) so large that the probability associated with their occurrence under H_0 is equal to or less than $\alpha = .01$.

The probability of obtaining 12 heads is $\dfrac{1}{4,096} = .00024$. Since $p = .00024$ is smaller than $\alpha = .01$, clearly the occurrence of 12 heads would be in the region of rejection.

The probability of obtaining either 11 *or* 12 heads is

$$\frac{1}{4,096} + \frac{12}{4,096} = \frac{13}{4,096} = .0032$$

Since $p = .0032$ is smaller than $\alpha = .01$, the occurrence of 11 heads would also be in the region of rejection.

The probability of obtaining 10 heads (or a value more extreme: 11 or 12 heads) is $\dfrac{1}{4,096} + \dfrac{12}{4,096} + \dfrac{66}{4,096} = \dfrac{79}{4,096} = .019$. Since $p = .019$ is larger than $\alpha = .01$, the occurrence of 10 heads would not be in the region of rejection. That is, if 10 or fewer heads turn up in our sample of 12 tosses we cannot reject H_0 at the $\alpha = .01$ level of significance.

vi. *Decision.* Suppose in our sample of tosses we obtain 11 heads. The probability associated with an occurrence as extreme as this one is $p = .0032$. Inasmuch as this p is smaller than our previously set level of significance ($\alpha = .01$), our decision is to reject H_0 in favor of H_1. We conclude that the coin is biased to land head up.

This chapter has discussed the procedure for making a decision as to whether a particular hypothesis, as operationally defined, should be accepted or rejected in terms of the information yielded by the research. Chapter 3 completes the general discussion by going into the question of how one may choose the most appropriate statistical test for use with one's research data. (This choice is step ii in the procedure outlined above.) The discussion in Chap. 3 clarifies the conditions under which the parametric tests are optimum and indicates the conditions under which nonparametric tests are more appropriate.

The reader who wishes to gain a more comprehensive or fundamental

EXAMPLE 17

understanding of the topics summarized in bare outline in the present chapter may refer to Dixon and Massey (1950, chap. 14) for an unusually clear introductory discussion of power functions and of the two types of errors, and to Anderson and Bancroft (1952, chap. 11) or Mood (1950, chap. 12) for more advanced discussions of the theory of testing hypotheses.

CHAPTER 3

CHOOSING AN APPROPRIATE STATISTICAL TEST

When alternative statistical tests are available for a given research design, as is very often the case, it is necessary to employ some rationale for choosing among them. In Chap. 2 we presented one criterion to use in choosing among alternative statistical tests: the criterion of power. In this chapter other criteria will be presented.

The reader will remember that the *power* of a statistical analysis is partly a function of the statistical test employed in the analysis. A statistical test is a good one if it has a small probability of rejecting H_0 when H_0 is true, but a large probability of rejecting H_0 when H_0 is false. Suppose we find two statistical tests, A and B, which have the same probability of rejecting H_0 when it is true. It might seem that we should simply select the one that has the larger probability of rejecting H_0 when it is false.

However, there are considerations other than power which enter into the choice of a statistical test. In this choice we must consider the manner in which the sample of scores was drawn, the nature of the population from which the sample was drawn, and the kind of measurement or scaling which was employed in the operational definitions of the variables involved, i.e., in the scores. All these matters enter into determining which statistical test is optimum or most appropriate for analyzing a particular set of research data.

THE STATISTICAL MODEL

When we have asserted the nature of the population and the manner of sampling, we have established a statistical model. Associated with every statistical test is a model and a measurement requirement; the test is valid under certain conditions, and the model and the measurement requirement specify those conditions. Sometimes we are able to test whether the conditions of a particular statistical model are met, but more often we have to *assume* that they are met. Thus the conditions of the statistical model of a test are often called the "assumptions" of the test. All decisions arrived at by the use of any statistical test must carry with them this qualification: "If the model used was correct, and if the measurement requirement was satisfied, then . . . ."

It is obvious that the fewer or weaker are the assumptions that define a particular model, the less qualifying we need to do about our decision arrived at by the statistical test associated with that model. That is, the fewer or weaker are the assumptions, the more general are the conclusions.

However, the most powerful tests are those which have the strongest or most extensive assumptions. The parametric tests, for example, the t or F tests, have a variety of strong assumptions underlying their use. When those assumptions are valid, these tests are the most likely of all tests to reject H_0 when H_0 is false. That is, when research data may appropriately be analyzed by a parametric test, that test will be more powerful than any other in rejecting H_0 when it is false. Notice, however, the requirement that the research data must be appropriate for the test. What constitutes such appropriateness? What are the conditions that are associated with the statistical model and the measurement requirement underlying, say, the t test? The conditions which must be satisfied to make the t test the most powerful one, and in fact before any confidence can be placed in any probability statement obtained by the use of the t test, are at least these:

1. The observations must be independent. That is, the selection of any one case from the population for inclusion in the sample must not bias the chances of any other case for inclusion, and the score which is assigned to any case must not bias the score which is assigned to any other case.

2. The observations must be drawn from normally distributed populations.

3. These populations must have the same variance (or, in special cases, they must have a known ratio of variances).

4. The variables involved must have been measured in *at least* an interval scale, so that it is possible to use the operations of arithmetic (adding, dividing, finding means, etc.) on the scores.

In the case of the analysis of variance (the F test), another condition is added to those already given:

5. The means of these normal and homoscedastic populations must be linear combinations of effects due to columns and/or rows. That is, the effects must be additive.

All the above conditions [except (4), which states the measurement requirement] are elements of the parametric statistical model. With the possible exception of the assumption of homoscedasticity (equal variances), these conditions are ordinarily not tested in the course of the performance of a statistical analysis. Rather, they are presumptions which are accepted, and their truth or falsity determines the meaningfulness of the probability statement arrived at by the parametric test.

When we have reason to believe that these conditions are met in the data under analysis, then we should certainly choose a parametric statistical test, such as t or F, for analyzing those data. Such a choice is optimum because the parametric test will be most powerful for rejecting H_0 when it should be rejected.

But what if these conditions are not met? What happens when the population is *not* normally distributed? What happens when the measurement is *not* so strong as an interval scale? What happens when the populations are *not* equal in variance?

When the assumptions constituting the statistical model for a test are in fact not met, or when the measurement is not of the required strength, then it is difficult if not impossible to say what is really the power of the test. It is even difficult to estimate the extent to which a probability statement about the hypothesis in question is meaningful when that probability statement results from the unacceptable application of a test. Although some empirical evidence has been gathered to show that *slight* deviations in meeting the assumptions underlying parametric tests may not have *radical* effects on the obtained probability figure, there is as yet no general agreement as to what constitutes a "slight" deviation.

POWER-EFFICIENCY

We have already noticed that the fewer or weaker are the assumptions that constitute a particular model, the more general are the conclusions derived from the application of the statistical test associated with that model but the less powerful is the test of H_0. This assertion is generally true for any given sample size. But it may not be true in the comparison of two statistical tests which are applied to two samples of unequal size. That is, if $N = 30$ in both instances, test A may be more powerful than test B. But the same test B may be more powerful with $N = 30$ than is test A with $N = 20$. In other words, we can avoid the dilemma of having to choose between power and generality by selecting a statistical test which has broad generality and then increasing its power to that of the most powerful test available by enlarging the size of the sample.

The concept of *power-efficiency* is concerned with the amount of increase in sample size which is necessary to make test B as powerful as test A. If test A is the most powerful known test of its type (when used with data which meet its conditions), and if test B is another test for the same research design which is just as powerful with N_b cases as is test A with

N_a cases, then

$$\text{Power-efficiency of test } B = (100) \frac{N_a}{N_b} \text{ per cent}$$

For example, if test B requires a sample of $N = 25$ cases to have the same power as test A has with $N = 20$ cases, then test B has power-efficiency of $(100)\frac{20}{25}$ per cent, i.e., its power-efficiency is 80 per cent. A power-efficiency of 80 per cent means that in order to equate the power of test A and test B (when all the conditions of both tests are met, and when test A is the more powerful) we need to draw 10 cases for test B for every 8 cases drawn for test A.

Thus we can avoid having to meet some of the assumptions of the most powerful tests, the parametric tests, without losing power by simply choosing a different test and drawing a larger N. In other words, by choosing another statistical test with fewer assumptions in its model and thus with greater generality than the t and F tests, and by enlarging our N, we can avoid having to make assumptions 2, 3, and 5 above, and still retain equivalent power to reject H_0.

Two other conditions, 1 and 4 above, underlie parametric statistical tests. Assumption 1, that the scores are independently drawn from the population, is an assumption which underlies all statistical tests, parametric or nonparametric. But assumption 4, which concerns the strength of measurement required for parametric tests—measurement must be at least in an interval scale—is not shared by all statistical tests. Different tests require measurement of different strengths. In order to understand the measurement requirements of the various statistical tests, the reader should be conversant with some of the basic notions in the theory of measurement. The discussion of measurement which occupies the next few pages gives the required information.

MEASUREMENT

When a physical scientist talks about measurement, he usually means the assigning of numbers to observations in such a way that the numbers are amenable to analysis by manipulation or operation according to certain rules. This analysis by manipulation will reveal new information about the objects being measured. In other words, the relation between the things being observed and the numbers assigned to the observations is so direct that by manipulating the numbers the physical scientist obtains new information about the things. For example, he may determine how much a homogeneous mass of material would weigh if cut in half by simply dividing its weight by 2.

The social scientist, taking physics as his model, usually attempts to do likewise in his scoring or measurement of social variables. But in his scaling the social scientist very often overlooks a fundamental fact in measurement theory. He overlooks the fact that in order for him to be able to make certain operations with numbers that have been assigned to observations, the structure of his method of mapping numbers (assigning scores) to observations must be *isomorphic* to some numerical structure which includes these operations. If two systems are isomorphic, their structures are the same in the relations and operations they allow.

For example, if a researcher collects data made up of numerical scores and then manipulates these scores by, say, adding and dividing (which are necessary operations in finding means and standard deviations), he is assuming that the structure of his measurement is isomorphic to that numerical structure known as arithmetic. That is, he is assuming that he has attained a high level of measurement.

The theory of measurement consists of a set of separate or distinct theories, each concerning a distinct *level* of measurement. The operations allowable on a given set of scores are dependent on the level of measurement achieved. Here we will discuss four levels of measurement— nominal, ordinal, interval, and ratio—and will discuss the operations and thus the statistics and statistical tests that are permitted with each level.

The Nominal or Classificatory Scale

Definition. Measurement at its weakest level exists when numbers or other symbols are used simply to classify an object, person, or characteristic. When numbers or other symbols are used to identify the groups to which various objects belong, these numbers or symbols constitute a nominal or classificatory scale.

Examples. The psychiatric system of diagnostic groups constitutes a nominal scale. When a diagnostician identifies a person as "schizophrenic," "paranoid," "manic-depressive," or "psychoneurotic," he is using a symbol to represent the class of persons to which this person belongs, and thus he is using nominal scaling.

The numbers on automobile license plates constitute a nominal scale. If the assignment of plate numbers is purely arbitrary, then each plated car is a member of a unique subclass. But if, as is common in the United States, a certain number or letter on the license plate indicates the county in which the car owner resides, then each subclass in the nominal scale consists of a group of entities: all owners residing in the same county. Here the assignment of numbers must be such that the same number (or letter) is given to all persons residing in the same county and that different numbers (or letters) are given to people residing in different counties. That is, the number or letter on the license plate must clearly indicate to which of a set of mutually exclusive subclasses the owner belongs.

Numbers on football jerseys and social-security numbers are other examples of the use of numbers in nominal scaling.

Formal properties. All scales have certain formal properties. These properties provide fairly exact definitions of the scale's characteristics, more exact definitions than we can give in verbal terms. These properties may be formulated more abstractly than we have done here by a set of axioms which specify the operations of scaling and the relations among the objects that have been scaled.

In a nominal scale, the scaling operation is partitioning a given class into a set of mutually exclusive subclasses. The only relation involved is that of *equivalence*. That is, the members of any one subclass must be equivalent in the property being scaled. This relation is symbolized by the familiar sign: $=$. The equivalence relation is reflexive, symmetrical, and transitive.[1]

Admissible operations. Since in any nominal scale the classification may be equally well represented by any set of symbols, the nominal scale is said to be "unique up to a one-to-one transformation." The symbols designating the various subclasses in the scale may be interchanged, if this is done consistently and completely. For example, when new license plates are issued, the license number which formerly stood for one county can be interchanged with that which had stood for another county. Nominal scaling would be preserved if this change-over were performed consistently and thoroughly in the issuing of all license plates. Such one-to-one transformations are sometimes called "the symmetric group of transformations."

Since the symbols which designate the various groups on a nominal scale may be interchanged without altering the essential information in the scale, the only kinds of admissible descriptive statistics are those which would be unchanged by such a transformation: the mode, frequency counts, etc. Under certain conditions, we can test hypotheses regarding the distribution of cases among categories by using the nonparametric statistical test, χ^2, or by using a test based on the binomial expansion. These tests are appropriate for nominal data because they focus on frequencies in categories, i.e., on enumerative data. The most common measure of association for nominal data is the contingency coefficient, C, a nonparametric statistic.

The Ordinal or Ranking Scale

Definition. It may happen that the objects in one category of a scale are not just different from the objects in other categories of that scale,

---

[1] *Reflexive:* $x = x$ for all values of x. *Symmetrical:* if $x = y$, then $y = x$. *Transitive:* if $x = y$ and $y = z$, then $x = z$.

but that they stand in some kind of *relation* to them. Typical relations among classes are: higher, more preferred, more difficult, more disturbed, more mature, etc. Such relations may be designated by the carat ($>$) which, in general, means "greater than." In reference to particular scales, $>$ may be used to designate *is preferred to, is higher than, is more difficult than*, etc. Its specific meaning depends on the nature of the relation that defines the scale.

Given a group of equivalence classes (i.e., given a nominal scale), if the relation $>$ holds between some but not all pairs of classes, we have a *partially ordered scale*. If the relation $>$ holds for all pairs of classes so that a complete rank ordering of classes arises, we have an *ordinal scale*.

Examples. Socioeconomic status, as conceived by Warner and his associates,[1] constitutes an ordinal scale. In prestige or social acceptability, all members of the upper middle class are higher than ($>$) all members of the lower middle class. The lower middles, in turn, are higher than the upper lowers. The $=$ relation holds among members of the same class, and the $>$ relation holds between any pair of classes.

The system of grades in the military services is another example of an ordinal scale. Sergeant $>$ corporal $>$ private.

Many personality inventories and tests of ability or aptitude result in scores which have the strength of ranks. Although the scores may appear to be more precise than ranks, generally these scales do not meet the requirements of any higher level of measurement and may properly be viewed as ordinal.

Formal properties. Axiomatically, the fundamental difference between a nominal and an ordinal scale is that the ordinal scale incorporates not only the relation of equivalence ($=$) but also the relation "greater than" ($>$). The latter relation is irreflexive, asymmetrical, and transitive.[2]

Admissible operations. Since any order-preserving transformation does not change the information contained in an ordinal scale, the scale is said to be "unique up to a monotonic transformation." That is, it does not matter what numbers we give to a pair of classes or to members of those classes, just as long as we give a higher number to the members of the class which is "greater" or "more preferred." (Of course, one may use the lower numbers for the "more preferred" grades. Thus we usually refer to excellent performance as "first-class," and to progressively inferior performances as "second-class" and "third-class." So long as we are consistent, it does not matter whether higher or lower numbers are used to denote "greater" or "more preferred.")

[1] Warner, W. L., Meeker, M., and Eells, K. 1949. *Social class in America.* New York: Science Research Associates.

[2] *Irreflexive:* it is not true for any x that $x > x$. *Asymmetrical:* if $x > y$, then $y \not> x$. *Transitive:* if $x > y$ and $y > z$, then $x > z$.

For example, a corporal in the army wears two stripes on his sleeve and a sergeant wears three. These insignia denote that sergeant > corporal. This relation would be equally well expressed if the corporal wore four stripes and the sergeant wore seven. That is, a transformation which does not change the order of the classes is completely admissible because *it does not involve any loss of information*. Any or all the numbers applied to classes in an ordinal scale may be changed in any fashion which does not alter the ordering (ranking) of the objects.

The statistic most appropriate for describing the central tendency of scores in an ordinal scale is the median, since the median is not affected by changes of any scores which are above or below it as long as the number of scores above and below remains the same. With ordinal scaling, hypotheses can be tested by using that large group of nonparametric statistical tests which are sometimes called "order statistics" or "ranking statistics." Correlation coefficients based on rankings (e.g., the Spearman r_S or the Kendall τ) are appropriate.

The only assumption made by some ranking tests is that the scores we observe are drawn from an underlying continuous distribution. Parametric tests also make this assumption. An underlying continuous variate is one that is not restricted to having only isolated values. It may have any value in a certain interval. A discrete variate, on the other hand, is one which can take on only a finite number of values; a continuous variate is one which can (but may not) take on a continuous infinity of values.

For some nonparametric techniques which require ordinal measurement, the requirement is that there be a continuum *underlying* the observed scores. The actual scores we observe may fall into discrete categories. For example, the actual scores may be either "pass" or "fail" on a particular item. We may well assume that underlying such a dichotomy there is a continuum of possible results. That is, some individuals who were categorized as failing may have been closer to passing than were others who were categorized as failing. Similarly, some passed only minimally, whereas others passed with ease and dispatch. The assumption is that "pass" and "fail" represent a continuum dichotomized into two intervals.

Similarly, in matters of opinion those who are classified as "agree" and "disagree" may be thought to fall on a continuum. Some who score as "agree" are actually not very concerned with the issue, whereas others are strongly convinced of their position. Those who "disagree" include those who are only mildly in disagreement as well as die-hard opponents.

Frequently the grossness of our measuring devices obscures the underlying continuity that may exist. If a variate is truly continuously distributed, then the probability of a tie is zero. However, tied scores frequently occur. Tied scores are almost invariably a reflection of the lack of sensitivity of our measuring instruments, which fail to distinguish the small differences which really exist between the tied observations. Therefore even when ties are observed it may not be unreasonable to assume that a continuous distribution underlies our gross measures.

At the risk of being excessively repetitious, the writer wishes to emphasize here that parametric statistical tests, which use means and standard deviations (i.e., which require the operations of arithmetic on the original scores), ought not to be used with data in an ordinal scale. The properties of an ordinal scale are *not* isomorphic to the numerical system known as arithmetic. When only the rank order of scores is known, means and standard deviations found on the scores themselves are *in error* to the extent that the successive intervals (distances between classes) on the scale are not equal. When parametric techniques of statistical inference are used with such data, any decisions about hypotheses are doubtful. Probability statements derived from the application of parametric statistical tests to ordinal data are in error to the extent that the structure of the method of collecting the data is not isomorphic to arithmetic. Inasmuch as most of the measurements made by behavioral scientists culminate in ordinal scales (this seems to be the case except in the field of psychophysics, and possibly in the use of a few carefully standardized tests), this point deserves strong emphasis.

Since this book is addressed to the behavioral scientist, and since the scales used by behavioral scientists typically are at best no stronger than ordinal, the major portion of this book is devoted to those methods which are appropriate for testing hypotheses with data measured in an ordinal scale. These methods, which also have much less circumscribing or restrictive assumptions in their statistical models than have parametric tests, make up the bulk of the nonparametric tests.

The Interval Scale

Definition. When a scale has all the characteristics of an ordinal scale, and when in addition the *distances* between any two numbers on the scale are of known size, then measurement considerably stronger than ordinality has been achieved. In such a case measurement has been achieved in the sense of an interval scale. That is, if our mapping of several classes of objects is so precise that we know just how large are the intervals (distances) between all objects on the scale, then we have achieved interval measurement. An interval scale is characterized by a common and constant unit of measurement which assigns a real number to all pairs of objects in the ordered set. In this sort of measurement, the ratio of any two intervals is independent of the unit of measurement and of the zero point. In an interval scale, the zero point and the unit of measurement are arbitrary.

Examples. We measure temperature on an interval scale. In fact, two different scales—centigrade and Fahrenheit—are commonly used. The unit of measurement and the zero point in measuring temperature are arbitrary; they are different for the two scales. However, both scales contain the same amount and the same kind of information. This is the case because they are linearly related. That is, a reading on one scale can be transformed to the equivalent reading on the other by the linear transformation

$$F = \tfrac{9}{5}C + 32$$

where F = number of degrees on Fahrenheit scale
   C = number of degrees on centrigrade scale

It can be shown that the ratios of temperature differences (intervals) are independent of the unit of measurement and of the zero point. For instance, "freezing" occurs at 0 degrees on the centigrade scale, and "boiling" occurs at 100 degrees. On the Fahrenheit scale, "freezing" occurs at 32 degrees and "boiling" at 212 degrees. Some other readings of the same temperature on the two scales are:

Centigrade	0	10	30	100
Fahrenheit	32	50	86	212

Notice that the ratio of the *differences* between temperature readings on one scale is equal to the ratio between the equivalent differences on the other scale. For example, on the centigrade scale the ratio of the differences between 30 and 10, and 10 and 0, is $\dfrac{30 - 10}{10 - 0} = 2$. For the comparable readings on the Fahrenheit scale, the ratio is $\dfrac{86 - 50}{50 - 32} = 2$. The ratio is the same in both cases: 2. In an interval scale, in other words, the ratio of any two intervals is independent of the unit used and of the zero point, both of which are arbitrary.

Most behavioral scientists aspire to create interval scales, and on infrequent occasions they succeed. Usually, however, what is taken for success comes because of the untested assumptions the scale maker is willing to make. One frequent assumption is that the variable being scaled is normally distributed in the individuals being tested. Having made this assumption, the scale maker manipulates the units of the scale until the *assumed* normal distribution is recovered from the individuals' scores. This procedure, of course, is only as good as the intuition of the investigator when he hits upon the distribution to assume.

Another assumption which is often made in order to create an apparent interval scale is the assumption that the person's answer of "yes" on any one item is exactly equivalent to his answering affirmatively on any other item. This assumption is made in order to satisfy the requirement that an interval scale have a common and constant unit of measurement. In ability or aptitude scales, the equivalent assumption is that giving the correct answer to any one item is exactly equivalent (in amount of ability shown) to giving the correct answer to any other item.

Formal properties. Axiomatically, it can be shown that the operations and relations which give rise to the structure of an interval scale are such that the *differences* in the scale are isomorphic to the structure of arithmetic. Numbers may be associated with the positions of the objects on an interval scale so that the operations of arithmetic may be meaningfully performed on the *differences* between these numbers.

In constructing an interval scale, one must not only be able to specify equivalences, as in a nominal scale, and greater-than relations, as in an ordinal scale, but one must also be able to specify the ratio of any two intervals.

Admissible operations. Any change in the numbers associated with the positions of the objects measured in an interval scale must preserve not only the ordering of the objects but also the relative differences between the objects. That is, the interval scale is "unique up to a linear transformation." Thus the information yielded by the scale is not affected if each number is multiplied by a positive constant and then a constant is added to this product, that is, $f(x) = ax + b$. (In the temperature example, $a = \frac{9}{5}$ and $b = 32$.)

We have already noticed that the zero point in an interval scale is arbitrary. This is inherent in the fact that the scale is subject to transformations which consist of adding a constant to the numbers making up the scale.

The interval scale is the first truly *quantitative* scale that we have encountered. All the common parametric statistics (means, standard deviations, Pearson correlations, etc.) are applicable to data in an interval scale, as are the common parametric statistical tests (t test, F test, etc.). If measurement in the sense of an interval scale has in fact been achieved, and if all of the assumptions in the statistical model (given on page 19) are adequately met, then the researcher *should* utilize parametric statistical tests. In such a case, nonparametric methods usually would not take advantage of all the information contained in the research data.

The Ratio Scale

Definition. When a scale has all the characteristics of an interval scale and in addition has a true zero point as its origin, it is called a ratio scale. In a ratio scale, the ratio of any two scale points is independent of the unit of measurement.

Example. We measure mass or weight in a ratio scale. The scale of ounces and pounds has a true zero point. So does the scale of grams. The ratio between any two weights is independent of the unit of measurement. For example, if we determine the weights of two different objects not only in pounds but also in grams, we would find that the ratio of the two pound weights is identical to the ratio of the two gram weights.

Formal properties. The operations and relations which give rise to the numerical values in a ratio scale are such that the scale is isomorphic to the structure of arithmetic. Therefore the operations of arithmetic are permissible on the numerical values assigned to the objects themselves, as well as on the intervals between numbers as is the case in the interval scale.

Ratio scales, most commonly encountered in the physical sciences, are achieved only when all four of these relations are operationally possible to attain: (a) equivalence, (b) greater than, (c) known ratio of any two intervals, and (d) known ratio of any two scale values.

Admissible operations. The numbers associated with the ratio scale values are "true" numbers with a true zero; only the unit of measurement is arbitrary. Thus the ratio scale is "unique up to multiplication by a positive constant." That is, the ratios between any two numbers are preserved when the scale values are all multiplied by a positive constant, and thus such a transformation does not alter the information contained in the scale.

Any statistical test is usable when ratio measurement has been achieved. In addition to using those previously mentioned as being appropriate for use with data in interval scales, with ratio scales one may use such statistics as the geometric mean and the coefficient of variation—statistics which require knowledge of the true zero point.

Summary

Measurement is the process of mapping or assigning numbers to objects or observations. The kind of measurement which is achieved is a function of the rules under which the numbers were assigned. The operations and relations employed in obtaining the scores define and limit the manipulations and operations which are permissible in handling the scores; the manipulations and operations must be those of the numerical structure to which the measurement is isomorphic.

Four of the most general scales were discussed: the nominal, ordinal, interval, and ratio scales. Nominal and ordinal measurement are the most common types achieved in the behavioral sciences. Data measured by either nominal or ordinal scales should be analyzed by the nonparametric methods.   Data measured in interval or ratio scales may be analyzed by parametric methods, if the assumptions of the parametric statistical model are tenable.

Table 3.1 summarizes the information in our discussion of various levels of measurement and of the kinds of statistics and statistical tests which are appropriate to each level when the assumptions of the tests' statistical models are satisfied.

TABLE 3.1. FOUR LEVELS OF MEASUREMENT AND THE STATISTICS APPROPRIATE TO EACH LEVEL

Scale	Defining relations	Examples of appropriate statistics	Appropriate statistical tests
Nominal	(1) Equivalence	Mode Frequency Contingency coefficient	Nonparametric statistical tests
Ordinal	(1) Equivalence (2) Greater than	Median Percentile Spearman r_s Kendall τ Kendall W	
Interval	(1) Equivalence (2) Greater than (3) Known ratio of any two intervals	Mean Standard deviation Pearson product-moment correlation Multiple product-moment correlation	Nonparametric and parametric statistical tests
Ratio	(1) Equivalence (2) Greater than (3) Known ratio of any two intervals (4) Known ratio of any two scale values	Geometric mean Coefficient of variation	

The reader may find other discussions of measurement in Bergman and Spence (1944), Coombs (1950; 1952), Davidson, Siegel, and Suppes (1955), Hempel (1952), Siegel (1956), and Stevens (1946; 1951).

PARAMETRIC AND NONPARAMETRIC STATISTICAL TESTS

A parametric statistical test is a test whose model specifies certain conditions (given on page 19) about the parameters of the population from which the research sample was drawn. Since these conditions are not ordinarily tested, they are assumed to hold. The meaningfulness of the results of a parametric test depends on the validity of these assumptions. Parametric tests also require that the scores under analysis result from measurement in the strength of at least an interval scale.

A nonparametric statistical test is a test whose model does not specify conditions about the parameters of the population from which the sample was drawn. Certain assumptions are associated with most nonparametric statistical tests, i.e., that the observations are independent and that the variable under study has underlying continuity, but these assumptions are fewer and much weaker than those associated with parametric tests. Moreover, nonparametric tests do not require measurement so strong as that required for the parametric tests; most nonparametric tests apply to data in an ordinal scale, and some apply also to data in a nominal scale.

In this chapter we have discussed the various criteria which should be considered in the choice of a statistical test for use in making a decision about a research hypothesis. These criteria are (a) the power of the test, (b) the applicability of the statistical model on which the test is based to the data of the research, (c) power-efficiency, and (d) the level of measurement achieved in the research. It has been stated that a parametric statistical test is most powerful when all the assumptions of its statistical model are met and when the variables under analysis are measured in at least an interval scale. However, even when all the parametric test's assumptions about the population and requirements about strength of measurement are satisfied, we know from the concept of power-efficiency that by increasing the sample size by an appropriate amount we can use a nonparametric test rather than the parametric one and yet retain the same power to reject H_0.

Because the power of any nonparametric test may be increased by simply increasing the size of N, and because behavioral scientists rarely achieve the sort of measurement which permits the meaningful use of parametric tests, nonparametric statistical tests deserve an increasingly prominent role in research in the behavioral sciences. This book presents a variety of nonparametric tests for the use of behavioral scientists. The use of parametric tests in research has been presented well in a variety of sources[1] and therefore we will not review those tests here.

In many of the nonparametric statistical tests to be presented, the data are changed from scores to ranks or even to signs. Such methods

---

[1] Among the many sources on parametric statistical tests, these are especially useful: Anderson and Bancroft (1952), Dixon and Massey (1951), Edwards (1954), Fisher (1934; 1935), McNemar (1955), Mood (1950), Snedecor (1946), Walker and Lev (1953).

may arouse the criticism that they "do not use all of the information in the sample" or that they "throw away information." The answer to this objection is contained in the answers to these questions: (a) Of the methods available, parametric and nonparametric, which uses the information in the sample most appropriately? (b) How important is it that the conclusions from the research apply generally rather than only to populations with normal distributions?

The answer to the first question depends on the level of measurement achieved in the research and on the researcher's knowledge of the population. If the measurement is weaker than that of an interval scale, by using parametric tests the researcher would "add information" and thereby create distortions which may be as great and as damaging as those introduced by the "throwing away of information" which occurs when scores are converted to ranks. Moreover, the assumptions which must be made to justify the use of parametric tests usually rest on conjecture and hope, for knowledge about the population parameters is almost invariably lacking. Finally, for some population distributions a nonparametric statistical test is clearly superior in power to a parametric one (Whitney, 1948).

The answer to the second question can be given only by the investigator as he considers the substantive aspects of the research problem.

The relevance of the discussion of this chapter to the choice between parametric and nonparametric statistical tests may be sharpened by the summary below, which lists the advantages and disadvantages of nonparametric statistical tests.

Advantages of Nonparametric Statistical Tests

1. Probability statements obtained from most nonparametric statistical tests are *exact* probabilities (except in the case of large samples, where excellent approximations are available), regardless of the shape of the population distribution from which the random sample was drawn. The accuracy of the probability statement does not depend on the shape of the population, although some nonparametric tests may assume identity of shape of two or more population distributions, and some others assume symmetrical population distributions. In certain cases, the nonparametric tests do assume that the underlying distribution is continuous, an assumption which they share with parametric tests.

2. If sample sizes as small as $N = 6$ are used, there is no alternative to using a nonparametric statistical test unless the nature of the population distribution is *known exactly*.

3. There are suitable nonparametric statistical tests for treating samples made up of observations from several *different* populations. None

of the parametric tests can handle such data without requiring us to make seemingly unrealistic assumptions.

4. Nonparametric statistical tests are available to treat data which are inherently in ranks as well as data whose seemingly numerical scores have the strength of ranks. That is, the researcher may only be able to say of his subjects that one has more or less of the characteristic than another, without being able to say *how much* more or less. For example, in studying such a variable as anxiety, we may be able to state that subject A is more anxious than subject B without knowing at all exactly how much more anxious A is. If data are inherently in ranks, or even if they can only be categorized as plus or minus (more or less, better or worse), they can be treated by nonparametric methods, whereas they cannot be treated by parametric methods unless precarious and perhaps unrealistic assumptions are made about the underlying distributions.

5. Nonparametric methods are available to treat data which are simply classificatory, i.e., are measured in a nominal scale. No parametric technique applies to such data.

6. Nonparametric statistical tests are typically much easier to learn and to apply than are parametric tests.

Disadvantages of Nonparametric Statistical Tests

1. If all the assumptions of the parametric statistical model are in fact met in the data, and if the measurement is of the required strength, then nonparametric statistical tests are wasteful of data. The degree of wastefulness is expressed by the power-efficiency of the nonparametric test. (It will be remembered that if a nonparametric statistical test has power-efficiency of, say, 90 per cent, this means that *where all the conditions of the parametric test are satisfied* the appropriate parametric test would be just as effective with a sample which is 10 per cent smaller than that used in the nonparametric analysis.)

2. There are as yet no nonparametric methods for testing interactions in the analysis of variance model, unless special assumptions are made about additivity. (Perhaps we should disregard this distinction because parametric statistical tests are also forced to make the assumption of additivity. However, the problem of higher-ordered interactions has yet to be dealt with in the literature of nonparametric methods.)[1]

Another objection that has been entered against nonparametric methods is that the tests and their accompanying tables of significant values have been widely scattered in various publications, many highly

[1] After this book had been set in type, a nonparametric test was presented which contributes to the solution of this problem. See Wilson, K. V. 1956. A distribution-free test of analysis of variance hypotheses. *Psychol. Bull.*, **53**, 96–101.

specialized, and they have therefore been comparatively inaccessible to the behavioral scientist. In preparing this book, the writer's intention has been to rob that objection of its force. This book attempts to present all the nonparametric techniques of statistical inference and measures of association that the behavioral scientist is likely to need, and it gives all of the tables necessary for the use of these techniques. Although this text is not exhaustive in its coverage of nonparametric tests—it could not be without being excessively redundant—enough tests are included in the chapters which follow to give the behavioral scientist wide latitude in choosing a nonparametric technique appropriate to his research design and useful for testing his research hypothesis.

CHAPTER 4

THE ONE-SAMPLE CASE

In this chapter we present those nonparametric statistical tests which may be used to test a hypothesis which calls for drawing just one sample. The tests tell us whether the particular sample could have come from some specified population. These tests are in contrast to the two-sample tests, which may be more familiar, which compare two samples and test whether it is likely that the two came from the same population.

The one-sample test is usually of the goodness-of-fit type. In the typical case, we draw a random sample and then test the hypothesis that this sample was drawn from a population with a specified distribution. Thus the one-sample test can answer questions like these: Is there a significant difference in location (central tendency) between the sample and the population? Is there a significant difference between the observed frequencies and the frequencies we would expect on the basis of some principle? Is there a significant difference between observed and expected proportions? Is it reasonable to believe that this sample has been drawn from a population of a specified shape or form (e.g., normal, rectangular)? Is it reasonable to believe that this sample is a random sample from some known population?

In the one-sample case a common parametric technique is to apply a t test to the difference between the observed (sample) mean and the expected (population) mean. The t test, strictly speaking, assumes that the observations or scores in the sample have come from a normally distributed population. The t test also requires that the observations be measured at least in an interval scale.

There are many sorts of data to which the t test may be inapplicable. The experimenter may find that (a) the assumptions and requirements of the t test are unrealistic for his data, (b) it is preferable to avoid making the assumptions of the t test and thus to gain greater generality for the conclusions, (c) the data of his research are inherently in ranks and thus not amenable to analysis by the t test, (d) the data may be simply classificatory or enumerative and thus not amenable to analysis by the t test, or (e) he is not interested only in differences in location but rather wishes to expose *any* kind of difference whatsoever. In such instances the experimenter may choose to use one of the one-sample nonparametric statistical tests presented in this chapter.

Four tests for the one-sample case will be presented. The chapter concludes with a comparison and contrast of these tests, which may aid the researcher in selecting the test best suited to his research hypothesis and to his data.

THE BINOMIAL TEST

Function and Rationale

There are populations which are conceived as consisting of only two classes. Examples of such classes are: male and female, literate and illiterate, member and nonmember, in-school and out-of-school, married and single, institutionalized and ambulatory. For such cases, all the possible observations from the population will fall into either one or the other of the two discrete classifications.

For any population of two classes, if we know that the proportion of cases in one class is P, then we know that the proportion in the other class must be $1 - P$. Usually the symbol Q is used for $1 - P$.

Although the value of P may vary from population to population, it is fixed for any one population. However, even if we know (or assume) the value of P for some population, we cannot expect that a random sample of observations from that population will contain exactly proportion P of cases in one class and proportion Q of cases in the other. Random effects of sampling will usually prevent the sample from exactly duplicating the population values of P and Q. For example, we may know from the official records that the voters in a certain county are evenly split between the Republican and Democratic parties in registration. But a random sample of the registered voters of that county might contain 47 per cent Democrats and 53 per cent Republicans, or even 56 per cent Democrats and 44 per cent Republicans. Such differences between the observed and the population values arise because of chance. Of course, small differences or deviations are more probable than large ones.

The binomial distribution is the sampling distribution of the proportions we might observe in random samples drawn from a two-class population. That is, it gives the various values which might occur under H_0. Here H_0 is the hypothesis that the population value is P. Therefore when the "scores" of a research are in two classes, the binomial distribution may be used to test H_0. The test is of the goodness-of-fit type. It tells us whether it is reasonable to believe that the proportions (or frequencies) we observe in our sample could have been drawn from a population having a specified value of P.

Method

The probability of obtaining x objects in one category and $N - x$ objects in the other category is given by

$$p(x) = \binom{N}{x} P^x Q^{N-x} \tag{4.1}$$

where P = proportion of cases expected in one of the categories

$Q = 1 - P$ = proportion of cases expected in the other category

$$\binom{N}{x} = \frac{N!}{x!\,(N-x)!} \quad *$$

A simple illustration will clarify formula (4.1). Suppose a fair die is rolled five times. What is the probability that exactly two of the rolls will show "six"? In this case, N = the number of rolls = 5; x = the number of sixes = 2; P = the expected proportion of sixes = $\frac{1}{6}$ (since the die is fair and therefore each side may be expected to show equally often); and $Q = 1 - P = \frac{5}{6}$. The probability that exactly two of the five rolls will show six is given by formula (4.1):

$$p(x) = \binom{N}{x} P^x Q^{N-x} \tag{4.1}$$

$$p(2) = \frac{5!}{2!\,3!} \left(\frac{1}{6}\right)^2 \left(\frac{5}{6}\right)^3$$

$$= .16$$

The application of the formula to the problem shows us that the probability of obtaining exactly two "sixes" when rolling a fair die five times is $p = .16$.

Now when we do research our question is usually *not* "What is the probability of obtaining *exactly* the values which were observed?" Rather, we usually ask, "What is the probability of obtaining the observed values or values even more extreme?" To answer questions of this type, the sampling distribution of the binomial is

$$\sum_{i=0}^{x} \binom{N}{i} P^i Q^{N-i} \tag{4.2}$$

---

* $N!$ is N factorial, which means $N(N-1)(N-2) \cdots (2)(1)$. For example, $4! = (4)(3)(2)(1) = 24$. Table S of the Appendix gives factorials for values through 20. Table T of the Appendix gives binomial coefficients, $\binom{N}{x}$, for values of N through 20.

In other words, we sum the probability of the observed value with the probabilities of values even more extreme.

Suppose now that we want to know the probability of obtaining two *or fewer* "sixes" when a fair die is rolled five times. Here again $N = 5$, $x = 2$, $P = \frac{1}{6}$, and $Q = \frac{5}{6}$. Now the probability of obtaining 2 or fewer "sixes" is $p(x \leq 2)$. The probability of obtaining 0 "sixes" is $p(0)$. The probability of obtaining 1 "six" is $p(1)$. The probability of obtaining 2 "sixes" is $p(2)$. We know from formula (4.2) above that

$$p(x \leq 2) = p(0) + p(1) + p(2)$$

That is, the probability of obtaining two or fewer "sixes" is the sum of the three probabilities mentioned above. If we use formula (4.1) to determine each of these probabilities, we have:

$$p(0) = \frac{5!}{0!\,5!} \left(\frac{1}{6}\right)^0 \left(\frac{5}{6}\right)^5 = .40$$

$$p(1) = \frac{5!}{1!\,4!} \left(\frac{1}{6}\right)^1 \left(\frac{5}{6}\right)^4 = .40$$

$$p(2) = \frac{5!}{2!\,3!} \left(\frac{1}{6}\right)^2 \left(\frac{5}{6}\right)^3 = .16$$

and thus
$$\begin{aligned} p(x \leq 2) &= p(0) + p(1) + p(2) \\ &= .40 + .40 + .16 \\ &= .96 \end{aligned}$$

We have determined that the probability under H_0 of obtaining two or fewer "sixes" when a fair die is rolled five times is $p = .96$.

Small samples. In the one-sample case, when a two-category class is used, a common situation is for P to equal $\frac{1}{2}$. Table D of the Appendix gives the one-tailed probabilities associated with the occurrence of various values as extreme as x under the null hypothesis that $P = Q = \frac{1}{2}$. When referring to Table D, let $x =$ the smaller of the observed frequencies. This table is useful when N is 25 or smaller. Its use obviates the necessity for using formula (4.2). When $P \neq Q$, formula (4.2) should be used. Table D gives the probabilities associated with the occurrence of various values as small as x for various N's (from 5 to 25). For example, suppose we observe 7 cases fall in one category while the other 3 fall in another. Here $N = 10$ and $x = 3$. Table D shows that the one-tailed probability of occurrence under H_0 of $x = 3$ or fewer when $N = 10$ is $p = .172$. (Notice that the decimal points have been omitted from the p's given in the body of the table.)

The p's given in Table D are one-tailed. A one-tailed test is used when we have predicted in advance which of the categories will contain the smaller number of cases. When the prediction is simply that the two frequencies will differ, a two-tailed test is used. For a two-tailed test, the p yielded by Table D is doubled. Thus for $N = 10$ and $x = 3$, the two-tailed probability associated with the occurrence under H_0 of such an extreme value of x is $p = 2(.172) = .344$.

The example which follows illustrates the use of the binomial test in a research where $P = Q = \frac{1}{2}$.

Example

In a study of the effects of stress,[1] an experimenter taught 18 college students 2 different methods to tie the same knot. Half of the subjects (randomly selected from the group of 18) learned method A first, and half learned method B first. Later—at midnight, after a 4-hour final examination—each subject was asked to tie the knot. The prediction was that stress would induce regression, i.e., that the subjects would revert to the first-learned method of tying the knot. Each subject was categorized according to whether he used the knot-tying method he learned first or the one he learned second, when asked to tie the knot under stress.

i. *Null Hypothesis.* H_0: $p_1 = p_2 = \frac{1}{2}$. That is, there is no difference between the probability of using the first-learned method under stress (p_1) and the probability of using the second-learned method under stress (p_2); any difference between the frequencies which may be observed is of such a magnitude that it might be expected in a sample from the population of possible results under H_0. H_1: $p_1 > p_2$.

ii. *Statistical Test.* The binomial test is chosen because the data are in two discrete categories and the design is of the one-sample type. Since methods A and B were randomly assigned to being first-learned and second-learned, there is no reason to think that the first-learned method would be preferred to the second-learned under H_0, and thus $P = Q = \frac{1}{2}$.

iii. *Significance Level.* Let $\alpha = .01$. $N =$ the number of cases $= 18$.

iv. *Sampling Distribution.* The sampling distribution is given in formula (4.2) above. However, when N is 25 or smaller, and when $P = Q = \frac{1}{2}$, Table D gives the probabilities associated with the occurrence under H_0 of observed values as small as x, and thus obviates the necessity for using the sampling distribution directly in the employment of this test.

v. *Rejection Region.* The region of rejection consists of all values

[1] Barthol, R. P., and Ku, Nani D. 1955. Specific regression under a nonrelated stress situation. *Amer. Psychologist*, **10**, 482. (Abstract)

of x (where x = the number of subjects who used the second-learned method under stress) which are so small that the probability associated with their occurrence under H_0 is equal to or less than α = .01. Since the direction of the difference was predicted in advance, the region of rejection is one-tailed.

vi. *Decision.* In the experiment, all but two of the subjects used the first-learned method when asked to tie the knot under stress (late at night after a long final examination). These data are shown in Table 4.1.

TABLE 4.1. KNOT-TYING METHOD CHOSEN UNDER STRESS

	Method chosen		Total
	First-learned	Second-learned	
Frequency	16	2	18

In this case, N = the number of independent observations = 18. x = the smaller frequency = 2. Table D shows that for N = 18, the probability associated with $x \leq 2$ is p = .001. Inasmuch as this p is smaller than α = .01, the decision is to reject H_0 in favor of H_1. We conclude that $p_1 > p_2$, that is, that persons under stress revert to the first-learned of two methods.

Large samples. Table D cannot be used when N is larger than 25. However, it can be shown that as N increases, the binomial distribution tends toward the normal distribution. This tendency is rapid when P is close to $\frac{1}{2}$, but slow when P is near 0 or 1. That is, the greater is the disparity between P and Q, the larger must be N before the approximation is usefully close. When P is near $\frac{1}{2}$, the approximation may be used for a statistical test for $N > 25$. When P is near 0 or 1, a rule of thumb is that NPQ must equal at least 9 before the statistical test based on the normal approximation is applicable. Within these limitations, the sampling distribution of x is approximately normal, with mean = NP and standard deviation = $\sqrt{NPQ}$, and therefore H_0 may be tested by

$$z = \frac{x - \mu_x}{\sigma_x} = \frac{x - NP}{\sqrt{NPQ}} \tag{4.3}$$

z is approximately normally distributed with zero mean and unit variance.

The approximation becomes an excellent one if a correction for continuity is incorporated. The correction is necessary because the normal distribution is for a continuous variable, whereas the binomial distribution involves a discrete variable. To correct for continuity, we regard the observed frequency x of formula (4.3) as occupying an interval, the lower limit of which is half a unit below the observed frequency while the upper limit is half a unit above the observed frequency. The correction for continuity consists of reducing, by .5, the difference between the observed value of x and the expected value, $\mu_x = NP$. Therefore when $x < \mu_x$ we add .5 to x, and when $x > \mu_x$ we subtract .5 from x. That is, the observed difference is reduced by .5. Thus z becomes

$$z = \frac{(x \pm .5) - NP}{\sqrt{NPQ}} \qquad (4.4)$$

where $x + .5$ is used when $x < NP$, and $x - .5$ is used when $x > NP$. The value of z obtained by the application of formula (4.4) may be considered to be normally distributed with zero mean and unit variance. Therefore the significance of an obtained z may be determined by reference to Table A of the Appendix. That is, Table A gives the one-tailed probability associated with the occurrence under H_0 of values as extreme as an observed z. (If a two-tailed test is required, the p yielded by Table A should be doubled.)

To show how good an approximation this is when $P = \frac{1}{2}$ even for $N < 25$, we can apply it to the knot-tying data discussed earlier. In that case, $N = 18$, $x = 2$, and $P = Q = \frac{1}{2}$. For these data, $x < NP$, that is, $2 < 9$, and, by formula (4.4),

$$z = \frac{(2 + .5) - (18)(.5)}{\sqrt{(18)(.5)(.5)}}$$
$$= -3.07$$

Table A shows that a z as extreme as -3.07 has a one-tailed probability associated with its occurrence under H_0 of $p = .0011$. This is essentially the same probability we found by the other analysis, which used a table of exact probabilities.

Summary of procedure. In brief, these are the steps in the use of the binomial test:

1. Determine $N =$ the total number of cases observed.

2. Determine the frequencies of the observed occurrences in each of the two categories.

3. The method of finding the probability of occurrence under H_0 of the observed values, or values even more extreme, varies:

    a. If N is 25 or smaller, and if $P = Q = \frac{1}{2}$, Table D gives the one-tailed probabilities under H_0 of various values as small as an observed x.

A one-tailed test is used when the researcher has predicted which category will have the smaller frequency. For a two-tailed test, double the p shown in Table D.

b. If $P \neq Q$, determine the probability of the occurrence under H_0 of the observed value of x or of an even more extreme value by substituting the observed values in formula (4.2). Table T is helpful in this computation; it gives binomial coefficients, $\binom{N}{x}$, for $N \leq 20$.

c. If N is larger than 25, and P close to $\frac{1}{2}$, test H_0 by using formula (4.4). Table A gives the probability associated with the occurrence under H_0 of values as large as an observed z yielded by that formula. Table A gives one-tailed p's; for a two-tailed test, double the p it yields.

If the p associated with the observed value of x or an even more extreme value is equal to or less than α, reject H_0.

Power-Efficiency

Inasmuch as there is no parametric technique applicable to data measured in a nominal scale, it would be meaningless to inquire about the power-efficiency of the binomial test when used with nominal data.

If a continuum is dichotomized and the binomial test used on the resulting data, that test may be wasteful of data. In such cases, the binomial test has power-efficiency (in the sense defined in Chap. 3) of 95 per cent for $N = 6$, decreasing to an eventual (asymptotic) efficiency of $\frac{2}{\pi} = 63$ per cent (Mood, 1954). However, if the data are basically dichotomous, even though the variable has an underlying continuous distribution, the binomial test may have no more powerful alternative.

References

For other discussions of the binomial test, the reader may turn to Clopper and Pearson (1934), David (1949, chaps. 3, 4), McNemar (1955, pp. 42–49), and Mood (1950, pp. 54–58).

THE χ^2 ONE-SAMPLE TEST

Function

Frequently research is undertaken in which the researcher is interested in the number of subjects, objects, or responses which fall in various categories. For example, a group of patients may be classified according to their preponderant type of Rorschach response, and the investigator may predict that certain types will be more frequent than others. Or children may be categorized according to their most frequent modes of play, to test the hypothesis that these modes will differ in frequency. Or persons may be categorized according to whether they are "in favor of," "indifferent to," or "opposed to" some statement of opinion, to enable the researcher to test the hypothesis that these responses will differ in frequency.

The χ^2 test is suitable for analyzing data like these. The number of categories may be two or more. The technique is of the goodness-of-fit type in that it may be used to test whether a significant difference exists between an *observed* number of objects or responses falling in each category and an *expected* number based on the null hypothesis.

Method

In order to be able to compare an observed with an expected group of frequencies, we must of course be able to state what frequencies would be expected. The null hypothesis states the proportion of objects falling in each of the categories in the presumed population. That is, from the null hypothesis we may deduce what are the expected frequencies. The χ^2 technique tests whether the observed frequencies are sufficiently close to the expected ones to be likely to have occurred under H_0.

The null hypothesis may be tested by

$$\chi^2 = \sum_{i=1}^{k} \frac{(O_i - E_i)^2}{E_i} \qquad (4.5)$$

where O_i = observed number of cases categorized in ith category

E_i = expected number of cases in ith category under H_0

$\sum_{i=1}^{k}$ directs one to sum over all (k) categories

Thus formula (4.5) directs one to sum over k categories the squared differences between each observed and expected frequency divided by the corresponding expected frequency.

If the agreement between the observed and expected frequencies is close, the differences $(O_i - E_i)$ will be small and consequently χ^2 will be small. If the divergence is large, however, the value of χ^2 as computed from formula (4.5) will also be large. Roughly speaking, the larger χ^2 is, the more likely it is that the observed frequencies did not come from the population on which the null hypothesis is based.

It can be shown that the sampling distribution of χ^2 under H_0, as computed from formula (4.5), follows the chi-square[1] distribution with

---

[1] To avoid confusion, the symbol χ^2 will be used for the quantity which is calculated from the observed data [using formula (4.5)] when a χ^2 test is performed. The words "chi square" will refer to a random variable which follows the chi-square distribution, certain values of which are shown in Table C.

$df = k - 1$. (df refers to degrees of freedom; these are discussed below.) Table C of the Appendix is taken from the sampling distribution of chi square, and gives certain critical values. At the top of each column in Table C are given the associated probabilities of occurrence (two-tailed) under H_0. The values in any column are the values of chi square which have the associated probability of occurrence under H_0 given at the top of that column. There is a different value of chi square for each df.

There are a number of different sampling distributions for chi square, one for each value of df. The size of df reflects the number of observations that are free to vary after certain restrictions have been placed on the data. These restrictions are not arbitrary, but rather are inherent in the organization of the data. For example, if the data for 50 cases are classified in two categories, then as soon as we know that, say, 35 cases fall in one category, we also know that 15 must fall in the other. For this example, $df = 1$, because with two categories and any fixed value of N, as soon as the number of cases in one category is ascertained then the number of cases in the other category is determined.

In general, for the one-sample case, when H_0 fully specifies the E_i's, $df = k - 1$, where k stands for the number of categories in the classification.

To use χ^2 in testing a hypothesis in the one-sample case, cast each observation into one of k cells. The total number of such observations should be N, the number of cases in your sample. That is, each observation must be independent of every other; thus one may not make several observations on the same person and count each as independent. To do so produces an "inflated N." For each of the k cells, the expected frequency must also be entered. If H_0 is that the proportion of cases in each category is the same, then $E_i = N/k$. With the various values of E_i and O_i known, one may compute the value of χ^2 by the application of formula (4.5). The significance of this obtained value of χ^2 may be determined by reference to Table C. If the probability associated with the occurrence under H_0 of the obtained χ^2 for $df = k - 1$ is equal to or less than the previously determined value of α, then H_0 may be rejected. If not, H_0 will be accepted.

Example

Horse-racing fans often maintain that in a race around a circular track significant advantages accrue to the horses in certain post positions. Any horse's post position is his assigned post in the starting line-up. Position 1 is closest to the rail on the inside of the track; position 8 is on the outside, farthest from the rail in an 8-horse race. We may test the effect of post position by analyzing the race results.

given according to post position, for the first month of racing in the 1955 season at a particular circular track.[1]

i. *Null Hypothesis.* H_0: there is no difference in the expected number of winners starting from each of the post positions, and any observed differences are merely chance variations to be expected in a random sample from the rectangular population where $f_1 = f_2 = \cdots = f_8$. H_1: the frequencies $f_1, f_2, \ldots, f_8$ are not all equal.

ii. *Statistical Test.* Since we are comparing the data from one sample with some presumed population, a one-sample test is appropriate. The χ^2 test is chosen because the hypothesis under test concerns a comparison of observed and expected frequencies in discrete categories. (The categories are the eight post positions.)[2]

iii. *Significance Level.* Let $\alpha = .01$. $N = 144$, the total number of winners in 18 days of racing.

iv. *Sampling Distribution.* The sampling distribution of χ^2, as computed from formula (4.5), follows the chi-square distribution with $df = k - 1$.

v. *Rejection Region.* H_0 will be rejected if the observed value of χ^2 is such that the probability associated with its occurrence under H_0 for $df = 7$ is equal to or less than $\alpha = .01$.

vi. *Decision.* The sample of 144 winners yielded the data shown in Table 4.2. The observed frequencies of wins are given in the

TABLE 4.2. WINS ACCRUED ON A CIRCULAR TRACK BY HORSES FROM EIGHT POST POSITIONS

	Post position								Total
	1	2	3	4	5	6	7	8	
	18	*18*	*18*	*18*	*18*	*18*	*18*	*18*	
No. of wins	29	19	18	25	17	10	15	11	144

center of each cell; the expected frequencies are given in italics in the corner of each cell. For example, 29 wins accrued to horses in position 1, whereas under H_0 only 18 wins would have been expected. And only 11 wins accrued to horses in position 8, whereas under H_0 18 would have been expected.

[1] The data are given in the *New York Post*, Aug. 30, 1955, p. 42.

[2] The χ^2 test may not be the most appropriate one for these data, since there seems to be some question of order involved, and the χ^2 test is insensitive to the effect of order. The example is presented because it illustrates the use and computation of the test. Later in this chapter we shall present one-sample tests which may be more appropriate for such data.

The computation of χ^2 is straightforward:

$$\chi^2 = \sum_{i=1}^{k} \frac{(O_i - E_i)^2}{E_i} \tag{4.5}$$

$$= \frac{(29 - 18)^2}{18} + \frac{(19 - 18)^2}{18} + \frac{(18 - 18)^2}{18} + \frac{(25 - 18)^2}{18}$$

$$+ \frac{(17 - 18)^2}{18} + \frac{(10 - 18)^2}{18} + \frac{(15 - 18)^2}{18} + \frac{(11 - 18)^2}{18}$$

$$= \tfrac{121}{18} + \tfrac{1}{18} + 0 + \tfrac{49}{18} + \tfrac{1}{18} + \tfrac{64}{18} + \tfrac{9}{18} + \tfrac{49}{18}$$

$$= 16.3$$

Table C shows that $\chi^2 \geq 16.3$ for $df = 7$ has probability of occurrence between $p = .05$ and $p = .02$. That is, $.05 > p > .02$. Inasmuch as that probability is larger than the previously set level of significance, $\alpha = .01$, we cannot reject H_0 at that significance level. We notice that the null hypothesis could have been rejected at $\alpha = .05$. It would seem that more data are necessary before any definite conclusions concerning H_1 can be made.

Small expected frequencies. When $df = 1$, that is, when $k = 2$, each *expected* frequency should be at least 5. When $df > 1$, that is, when $k > 2$, the χ^2 test for the one-sample case should not be used when more than 20 per cent of the expected frequencies are smaller than 5 or when any expected frequency is smaller than 1 (Cochran, 1954). Expected frequencies sometimes can be increased by combining adjacent categories. This is desirable only if combinations can meaningfully be made (and of course if there are more than two categories to begin with).

For example, a sample of persons may be categorized according to whether their response to a statement of opinion is *strongly support, support, indifferent, oppose,* or *strongly oppose.* To increase E_i's, adjacent categories could be combined, and the persons categorized as *support, indifferent,* or *oppose,* or possibly as *support, indifferent, oppose,* and *strongly oppose.*

If one starts with but two categories and has an expected frequency of less than 5, or if after combining adjacent categories one ends up with but two categories and still has an expected frequency of less than 5, then the binomial test (pages 36 to 42) should be used rather than the χ^2 test to determine the probability associated with the occurrence of the observed frequencies under H_0.

Summary of procedure. In this discussion of the method for using the χ^2 test in the one-sample case, we have shown that the procedure for using the test involves these steps:

1. Cast the observed frequencies in the k categories. The sum of the frequencies should be N, the number of independent observations.

2. From H_0, determine the expected frequencies (the E_i's) for each of the k cells. Where $k > 2$, if more than 20 per cent of the E_i's are smaller than 5, combine adjacent categories, where this is reasonable, thereby reducing the value of k and increasing the values of some of the E_i's. Where $k = 2$, the χ^2 test for the one-sample case may be used appropriately only if each expected frequency is 5 or larger.

3. Using formula (4.5), compute the value of χ^2.

4. Determine the value of df. $df = k - 1$.

5. By reference to Table C, determine the probability associated with the occurrence under H_0 of a value as large as the observed value of χ^2 for the observed value of df. If that p is equal to or less than α, reject H_0.

Power

The literature does not contain much information about the power function of the χ^2 test. Inasmuch as this test is most commonly used when we do not have a clear alternative available, we are usually not in a position to compute the exact power of the test.

When nominal measurement is used or when the data consist of frequencies in inherently discrete categories, then the notion of power-efficiency of the χ^2 test is meaningless, for in such cases there is no parametric test that is suitable. If the data are such that a parametric test is available, then the χ^2 test may be wasteful of information.

It should be noted that when $df > 1$, χ^2 tests are insensitive to the effects of order, and thus when a hypothesis takes order into account, χ^2 may not be the best test. For methods that strengthen the common χ^2 tests when H_0 is tested against specific alternatives, see Cochran (1954).

References

Useful discussions of this χ^2 test are contained in Cochran (1952; 1954), Dixon and Massey (1951, chap. 13), Lewis and Burke (1949), and McNemar (1955, chap. 13).

THE KOLMOGOROV-SMIRNOV ONE-SAMPLE TEST

Function and Rationale

The Kolmogorov-Smirnov one-sample test is a test of goodness of fit. That is, it is concerned with the degree of agreement between the distribution of a set of sample values (observed scores) and some specified theoretical distribution. It determines whether the scores in the sample can reasonably be thought to have come from a population having the theoretical distribution.

Briefly, the test involves specifying the *cumulative* frequency distribution which would occur under the theoretical distribution and comparing that with the observed cumulative frequency distribution. The theoretical distribution represents what would be expected under H_0. The point at which these two distributions, theoretical and observed, show the greatest divergence is determined. Reference to the sampling distribution indicates whether such a large divergence is likely on the basis of chance. That is, the sampling distribution indicates whether a divergence of the observed magnitude would probably occur if the observations were really a random sample from the theoretical distribution.

Method

Let $F_0(X)$ = a completely specified cumulative frequency distribution function, the theoretical cumulative distribution under H_0. That is, for any value of X, the value of $F_0(X)$ is the proportion of cases expected to have scores equal to or less than X.

And let $S_N(X)$ = the observed cumulative frequency distribution of a random sample of N observations. Where X is any possible score, $S_N(X) = k/N$, where k = the number of observations equal to or less than X.

Now under the null hypothesis that the sample has been drawn from the specified theoretical distribution, it is expected that for every value of X, $S_N(X)$ should be fairly close to $F_0(X)$. That is, under H_0 we would expect the differences between $S_N(X)$ and $F_0(X)$ to be small and within the limits of random errors. The Kolmogorov-Smirnov test focuses on the *largest* of the deviations. The largest value of $F_0(X) - S_N(X)$ is called the *maximum deviation*, D:

$$D = \text{maximum } |F_0(X) - S_N(X)| \tag{4.6}$$

The sampling distribution of D under H_0 is known. Table E of the Appendix gives certain critical values from that sampling distribution. Notice that the significance of a given value of D depends on N.

For example, suppose one found by formula (4.6) that $D = .325$ when $N = 15$. Table E shows that $D \geq .325$ has an associated probability of occurrence (two-tailed) between $p = .10$ and $.05$.

If N is over 35, one determines the critical values of D by the divisions indicated in Table E. For example, suppose a researcher uses $N = 43$ cases and sets $\alpha = .05$. Table E shows that any D equal to or greater than $\dfrac{1.36}{\sqrt{N}}$ will be significant. That is, any D, as defined by formula (4.6), which is equal to or greater than $\dfrac{1.36}{\sqrt{43}} = .207$ will be significant at the .05 level (two-tailed test).

Critical values for one-tailed tests have not as yet been adequately tabled. For a method of finding associated probabilities for one-tailed tests, the reader may refer to Birnbaum and Tingey (1951) and Goodman (1954, p. 166).

Example

Suppose a researcher were interested in confirming by experimental means the sociological observation that American Negroes seem to have a hierarchy of preferences among shades of skin color.[1] To test how systematic Negroes' skin-color preferences are, our fictitious researcher arranges to have a photograph taken of each of ten Negro subjects. The photographer develops these in such a way that he obtains five copies of each photograph, each copy differing slightly in darkness from the others, so that the five copies can reliably be ranked from darkest to lightest skin color. The picture showing the darkest skin color for any subject is ranked as 1, the next darkest as 2, and so on, the lightest being ranked as 5. Each subject is then offered a choice among the five prints of his own photograph. If skin shade is unimportant to the subjects, the photographs of each rank should be chosen equally often except for random differences. If skin shade is important, as we hypothesize, then the subjects should consistently favor one of the extreme ranks.

i. *Null Hypothesis.* H_0: there is no difference in the expected number of choices for each of the five ranks, and any observed differences are merely chance variations to be expected in a random sample from the rectangular population where $f_1 = f_2 = \cdots = f_5$. H_1: the frequencies $f_1, f_2, \ldots, f_5$ are not all equal.

ii. *Statistical Test.* The Kolmogorov-Smirnov one-sample test is chosen because the researcher wishes to compare an observed distribution of scores on an ordinal scale with a theoretical distribution.

iii. *Significance Level.* Let $\alpha = .01$. $N =$ the number of Negroes who served as subjects in the study $= 10$.

iv. *Sampling Distribution.* Various critical values of D from the sampling distribution are presented in Table E, together with their associated probabilities of occurrence under H_0.

v. *Rejection Region.* The region of rejection consists of all values of D [computed by formula (4.6)] which are so large that the probability associated with their occurrence under H_0 is equal to or less than $\alpha = .01$.

vi. *Decision.* In this hypothetical study, each Negro subject chooses one of five prints of the same photograph. Suppose one subject chooses print 2 (the next-to-darkest print), five subjects choose

[1] Warner, W. L., Buford, H. J., and Walter, A. A. 1941. *Color and human nature.* Washington: American Council on Education.

print 4 (the next-to-lightest print), and four choose print 5 (the lightest print). Table 4.3 shows these data and casts them in the form appropriate for applying the Kolmogorov-Smirnov one-sample test.

TABLE 4.3. HYPOTHETICAL SKIN-COLOR PREFERENCES OF
10 NEGRO SUBJECTS

	Rank of photo chosen (1 is darkest skin color)				
	1	2	3	4	5
f = number of subjects choosing that rank	0	1	0	5	4
$F_0(X)$ = theoretical cumulative distribution of choices under H_0	$\frac{1}{5}$	$\frac{2}{5}$	$\frac{3}{5}$	$\frac{4}{5}$	$\frac{5}{5}$
$S_{10}(X)$ = cumulative distribution of observed choices	$\frac{0}{10}$	$\frac{1}{10}$	$\frac{1}{10}$	$\frac{6}{10}$	$\frac{10}{10}$
$|F_0(X) - S_{10}(X)|$	$\frac{2}{10}$	$\frac{3}{10}$	$\frac{5}{10}$	$\frac{2}{10}$	0

Notice that $F_0(X)$ is the theoretical cumulative distribution under H_0, where H_0 is that each of the 5 prints would receive $\frac{1}{5}$ of the choices. $S_{10}(X)$ is the cumulative distribution of the observed choices of the 10 Negro subjects. The bottom row of Table 4.3 gives the absolute deviation of each sample value from its paired expected value. Thus the first absolute deviation is $\frac{2}{10}$, which is obtained by subtracting 0 from $\frac{1}{5}$.

Inspection of the bottom row of Table 4.3 quickly reveals that the D for these data is $\frac{5}{10}$, which is .500. Table E shows that for $N = 10$, $D \geq .500$ has an associated probability under H_0 of $p < .01$. Inasmuch as the p associated with the observed value of D is smaller than $\alpha = .01$, our decision in this fictitious study is to reject H_0 in favor of H_1. We conclude that our subjects show significant preferences among skin colors.

Summary of procedure. In the computation of the Kolmogorov-Smirnov test, these are the steps:

1. Specify the theoretical cumulative step function, i.e., the cumulative distribution expected under H_0.

2. Arrange the observed scores in a cumulative distribution, pairing each interval of $S_N(X)$ with the comparable interval of $F_0(X)$.

3. For each step on the cumulative distributions, subtract $S_N(X)$ from $F_0(X)$.

4. Using formula (4.6), find D.

5. Refer to Table E to find the probability (two-tailed) associated with the occurrence under H_0 of values as large as the observed value of D. If that p is equal to or less than α, reject H_0.

Power

The Kolmogorov-Smirnov one-sample test treats individual observations separately and thus, unlike the χ^2 test for one sample, need not lose information through the combining of categories. When samples are small, and therefore adjacent categories must be combined before χ^2 may properly be computed, the χ^2 test is definitely less powerful than the Kolmogorov-Smirnov test. Moreover, for very small samples the χ^2 test is not applicable at all, but the Kolmogorov-Smirnov test is. These facts suggest that the Kolmogorov-Smirnov test may in all cases be more powerful than its alternative, the χ^2 test.

A reanalysis by the χ^2 test of the data given in the example above will highlight the superior power of the Kolmogorov-Smirnov test. In the form in which the data are presented in Table 4.3, χ^2 could not be computed, because the expected frequencies are only 2 when $N = 10$ and $k = 5$. We must combine adjacent categories in order to increase the expected frequency per cell. By doing that we end up with the two-category breakdown shown in Table 4.4. Any subject's choice is simply classified as being for a *light* or a *dark* skin color; finer gradations must be ignored.

TABLE 4.4. HYPOTHETICAL SKIN-COLOR PREFERENCES OF 10 NEGRO SUBJECTS

	Skin color of photograph chosen		Total
	Dark (ranks 1, 2)	Light (ranks 3, 4, 5)	
Frequency of choice	1	9	10

For these data, χ^2 (uncorrected for continuity) $= 3.75$. Table C shows that the probability associated with the occurrence under H_0 of such a value when $df = k - 1 = 1$ is between .10 and .05. That is, $.10 > p > .05$. This value of p does not enable us to reject H_0 at the .01 level of significance.

Notice that the p we found by the Kolmogorov-Smirnov test is smaller than .01, while that found by the χ^2 test is larger than .05. This difference gives some indication of the superior power of the Kolmogorov-Smirnov test.

References

The reader may find other discussions of the Kolmogorov-Smirnov test in Birnbaum (1952; 1953), Birnbaum and Tingey (1951), Goodman (1954), and Massey (1951a).

THE ONE-SAMPLE RUNS TEST

Function and Rationale

If an experimenter wishes to arrive at some conclusion about a population by using the information contained in a sample from that population, then his sample must be a random one.   In recent years, several techniques have been developed to enable us to test the hypothesis that a sample is random.   These techniques are based on the *order* or *sequence* in which the individual scores or observations originally were obtained.

The technique to be presented here is based on the number of runs which a sample exhibits.   A *run* is defined as a succession of identical symbols which are followed and preceded by different symbols or by no symbols at all.

For example, suppose a series of plus or minus scores occurred in this order:

$$+ \; + \; - \; - \; - \; + \; - \; - \; - \; - \; - \; + \; + \; - \; +$$

This sample of scores begins with a run of 2 pluses.   A run of 3 minuses follows.   Then comes another run which consists of 1 plus.   It is followed by a run of 4 minuses, after which comes a run of 2 pluses, etc.   We can group these scores into runs by underlining and numbering each succession of identical symbols:

$$\underset{1}{\underline{+\,+}} \quad \underset{2}{\underline{-\,-\,-}} \quad \underset{3}{\underline{+}} \quad \underset{4}{\underline{-\,-\,-\,-}} \quad \underset{5}{\underline{+\,+}} \quad \underset{6}{\underline{-}} \quad \underset{7}{\underline{+}}$$

We observe 7 runs in all: r = number of runs = 7.

The total number of runs in a sample of any given size gives an indication of whether or not the sample is random.   If very few runs occur, a time trend or some bunching due to lack of independence is suggested. If a great many runs occur, systematic short-period cyclical fluctuations seem to be influencing the scores.

For example, suppose a coin were tossed 20 times and the following sequence of heads (H) and tails (T) was observed:

$$H \quad H \quad H \quad H \quad H \quad H \quad H \quad H \quad H \quad H \quad T \quad T \quad T \quad T \quad T \quad T \quad T \quad T \quad T \quad T$$

Only two runs occurred in 20 tosses.   This would seem to be too few for a "fair" coin (or a fair tosser!).   Some lack of independence in the events is suggested. On the other hand, suppose the following sequence occurred:

$$H \quad T \quad H \quad T \quad H \quad T \quad H \quad T \quad H \quad T \quad H \quad T \quad H \quad T \quad H \quad T \quad H \quad T \quad H \quad T$$

Here too many runs are observed. In this case, with $r = 20$ when $N = 20$, it would also seem reasonable to reject the hypothesis that the coin is "fair." Neither of the above sequences seems to be a random series of H's and T's.

Notice that our analysis, which is based on the *order* of the events, gives us information which is not indicated by the *frequency* of the events. In both of the above cases, 10 tails and 10 heads occurred. If the scores were analyzed according to their frequencies, e.g., by use of the χ^2 test or the binomial test, we would have no reason to suspect the "fairness" of the coin. It is only a runs test, focusing on the order of the events, which reveals the striking lack of randomness of the scores and thus the possible lack of "fairness" in the coin.

The sampling distribution of the values of r which we could expect from repeated random samples is known. Using this sampling distribution, we may decide whether a given observed sample has more or fewer runs than would probably occur in a random sample.

Method

Let $n_1 =$ the number of elements of one kind, and $n_2 =$ the number of elements of the other kind. That is, n_1 might be the number of heads and n_2 the number of tails; or n_1 might be the number of pluses and n_2 the number of minuses. $N =$ the total number of observed events $= n_1 + n_2$.

To use the one-sample runs test, first observe the n_1 and n_2 events in the sequence in which they occurred and determine the value of r, the number of runs.

Small samples. If both n_1 and n_2 are equal to or less than 20, then Table F of the Appendix gives the critical values of r under H_0 for $\alpha = .05$. These are critical values from the sampling distribution of r under H_0. If the observed value of r falls between the critical values, we accept H_0. If the observed value of r is equal to or more extreme than one of the critical values, we reject H_0.

Two tables are given: F_I and F_{II}. Table F_I gives values of r which are so *small* that the probability associated with their occurrence under H_0 is $p = .025$. Table F_{II} gives values of r which are so *large* that the probability associated with their occurrence under H_0 is $p = .025$.

Any observed value of r which is equal to or less than the value shown in Table F_I *or* is equal to or larger than the value shown in Table F_{II} is in the region of rejection for $\alpha = .05$.

For example, in the first tossing of the coin discussed above, we observed two runs: one run of 10 heads followed by one run of 10 tails. Here $n_1 = 10$, $n_2 = 10$, and $r = 2$. Table F shows that for these values of n_1 and n_2, a random sample would be expected to contain more than 6 runs but less than 16. Any observed r of 6 or less or of 16 or more is in the region of rejection for $\alpha = .05$. The observed $r = 2$ is smaller than 6, so at the .05 significance level we reject the null hypothesis that the coin is producing a random series of heads and tails.

If a one-tailed test is called for, i.e., if the direction of the deviation from randomness is predicted in advance, then only one of the two tables need be examined. If the prediction is that too few runs will be observed, Table F_I gives the critical values of r. If the observed r under such a one-tailed test is equal to or smaller than that shown in Table F_I, H_0 may be rejected at $\alpha = .025$. If the prediction is that too many runs will be observed, Table F_{II} gives the critical values of r which are significant at the .025 level.

For example, take the case of the second sequence of coin tosses reported above. Suppose we had predicted in advance, for some reason this writer cannot imagine, that this coin would produce too many runs. We observe that $r = 20$ for $n_1 = 10$ and $n_2 = 10$. Since our observed value of r is equal to or larger than that shown in Table F_{II}, we may reject H_0 at $\alpha = .025$, and conclude that the coin is "unfair" in the predicted direction.

Example for Small Samples

In a study of the dynamics of aggression in young children, the experimenter observed pairs of children in a controlled play situation.[1] Most of the 24 children who served as subjects in the study came from the same nursery school and thus played together daily. Since the experimenter was able to arrange to observe but two children on any day, she was concerned that biases might be introduced into the study by discussions between those children who had already served as subjects and those who were to serve later. If such discussions had any effect on the level of aggression in the play sessions, this effect might show up as a lack of randomness in the aggression scores in the order in which they were collected. After the study was completed, the randomness of the sequence of scores was tested by converting each child's aggression score to a plus or minus, depending on whether it fell above or below the group median, and then applying the one-sample runs test to the observed sequence of pluses and minuses.

[1] Siegel, Alberta E. 1955. The effect of film-mediated fantasy aggression on strength of aggressive drive in young children. Unpublished doctor's dissertation, Stanford University.

i. *Null Hypothesis.* H_0: the pluses and minuses occur in random order. H_1: the order of the pluses and minuses deviates from randomness.

ii. *Statistical Test.* Since the hypothesis concerns the randomness of a single sequence of observations, the one-sample runs test is chosen.

iii. *Significance Level.* Let $\alpha = .05$. $N =$ the number of subjects $= 24$. Since the scores will be characterized as plus or minus depending on whether they fall above or below the middlemost score in the group, $n_1 = 12$ and $n_2 = 12$.

iv. *Sampling Distribution.* Table F gives the critical values of r from the sampling distribution.

v. *Rejection Region.* Since H_1 does not predict the direction of the deviation from randomness, a two-tailed test is used. H_0 will be rejected at the .05 level of significance if the observed r is either equal to or less than the appropriate value in Table F_I or is equal to or larger than the appropriate value in Table F_{II}. For $n_1 = 12$ and $n_2 = 12$, Table F shows that the region of rejection consists of all r's of 7 or less and all r's of 19 or more.

vi. *Decision.* Table 4.5 shows the aggression scores for each child in the order in which those scores occurred. The median of this set of scores is 24.5. All scores falling below that median are designated

TABLE 4.5. AGGRESSION SCORES IN ORDER OF OCCURRENCE

Child	Score	Position of score with respect to median	Child	Score	Position of score with respect to median
1	31	+	13	15	−
2	23	−	14	18	−
3	36	+	15	78	+
4	43	+	16	24	−
5	51	+	17	13	−
6	44	+	18	27	+
7	12	−	19	86	+
8	26	+	20	61	+
9	43	+	21	13	−
10	75	+	22	7	−
11	2	−	23	6	−
12	3	−	24	8	−

as minus in Table 4.5; all above that median are designated as plus. From the column showing the sequence of +'s and −'s the reader can readily observe that 10 runs occurred in this series, that is, $r = 10$.

Reference to Table F reveals that $r = 10$ for $n_1 = 12$ and $n_2 = 12$ does not fall in the region of rejection, and thus our decision is that the null hypothesis that the sample of scores occurred in random order is acceptable.

Large samples. If either n_1 or n_2 is larger than 20, Table F cannot be used. For such large samples, a good approximation to the sampling distribution of r is the normal distribution, with

$$\text{Mean} = \mu_r = \frac{2n_1n_2}{n_1 + n_2} + 1$$

and   Standard deviation $= \sigma_r = \sqrt{\dfrac{2n_1n_2(2n_1n_2 - n_1 - n_2)}{(n_1 + n_2)^2(n_1 + n_2 - 1)}}$

Therefore, when either n_1 or n_2 is larger than 20, H_0 may be tested by

$$z = \frac{r - \mu_r}{\sigma_r} = \frac{r - \left(\dfrac{2n_1n_2}{n_1 + n_2} + 1\right)}{\sqrt{\dfrac{2n_1n_2(2n_1n_2 - n_1 - n_2)}{(n_1 + n_2)^2(n_1 + n_2 - 1)}}} \tag{4.7}$$

Since the values of z which are yielded by formula (4.7) under H_0 are approximately normally distributed with zero mean and unit variance, the significance of any observed value of z computed from this formula may be determined by reference to the normal curve table, Table A of the Appendix. That is, Table A gives the one-tailed probabilities associated with the occurrence under H_0 of values as extreme as an observed z.

The large-sample example which follows uses this normal curve approximation to the sampling distribution of r.

Example for Large Samples

The writer was interested in ascertaining whether the arrangement of men and women in the queue in front of the box office of a motion-picture theater was a random arrangement. The data were obtained by simply tallying the sex of each of a succession of 50 persons as they approached the box office.

i. *Null Hypothesis.* H_0: the order of males and females in the queue was random. H_1: the order of males and females was not random.

ii. *Statistical Test.* The one-sample runs test was chosen because the hypothesis concerns the randomness of a single group of events.

iii. *Significance Level.* Let $\alpha = .05$. $N = 50 =$ the number of persons observed. The values of n_1 and n_2 will be determined only after the data are collected.

iv. *Sampling Distribution.* For large samples, the values of z which are computed from formula (4.7) under H_0 are approximately normally distributed. Table A gives the one-tailed probability associated with the occurrence under H_0 of values as extreme as an observed z.

v. *Rejection Region.* Since H_1 does not predict the direction of the deviation from randomness, a two-tailed region of rejection is used. It consists of all values of z, as computed from formula (4.7), which are so extreme that the probability associated with their occurrence under H_0 is equal to or less than $\alpha = .05$. Thus the region of rejection includes all values of z equal to or more extreme than ± 1.96.

vi. *Decision.* The males (M) and females (F) were queued in front of the box office in the order shown in Table 4.6. The reader

TABLE 4.6. ORDER OF 30 MALES (M) AND 20 FEMALES (F) IN QUEUE
BEFORE THEATER BOX OFFICE
(Runs are indicated by underlining)

$$M \quad F \quad M \quad F \quad M\,M\,M \quad F\,F \quad M \quad F \quad M \quad F$$

$$M \quad F \quad M\,M\,M\,M \quad F \quad M \quad F \quad M \quad F \quad M\,M$$

$$F\,F\,F \quad M \quad F \quad M \quad F \quad M \quad F \quad M\,M \quad F$$

$$M\,M \quad F \quad M\,M\,M\,M \quad F \quad M \quad F \quad M\,M$$

will observe that there were 30 males and 20 females in this sample. By inspection of the data in Table 4.6, he may also readily determine that $r = 35 =$ the number of runs.

To determine whether $r \geq 35$ might readily have occurred under H_0, we compute the value of z as defined by formula (4.7). Let $n_1 =$ the number of males $= 30$, and $n_2 =$ the number of females $= 20$. Then

$$z = \frac{r - \left(\dfrac{2n_1n_2}{n_1 + n_2} + 1\right)}{\sqrt{\dfrac{2n_1n_2(2n_1n_2 - n_1 - n_2)}{(n_1 + n_2)^2(n_1 + n_2 - 1)}}} \tag{4.7}$$

$$= \frac{35 - \left(\dfrac{2(30)(20)}{30 + 20} + 1\right)}{\sqrt{\dfrac{2(30)(20)[2(30)(20) - 30 - 20]}{(30 + 20)^2(30 + 20 - 1)}}}$$

$$= 2.98$$

Table A shows that the probability of occurrence under H_0 of

$z \geq 2.98$ is $p = 2(.0014) = .0028$. (The probability is twice the p given in the table because a two-tailed test is called for.) Inasmuch as the probability associated with the observed occurrence, $p = .0028$, is less than the level of significance, $\alpha = .05$, our decision is to reject the null hypothesis in favor of the alternative hypothesis. We conclude that in the queue the order of males and females was not random.

Summary of procedure. These are the steps in the use of the one-sample runs test:

1. Arrange the n_1 and n_2 observations in their order of occurrence.
2. Count the number of runs, r.
3. Determine the probability under H_0 associated with a value as extreme as the observed value of r. If that probability is equal to or less than α, reject H_0. The technique for determining the value of p depends on the size of the n_1 and n_2 groups:

   a. If n_1 and n_2 are both 20 or less, refer to Table F. Table F_I gives the the value of r which is so small that its associated probability under H_0 is $p = .025$; Table F_{II} gives the value of r which is so large that its associated probability under H_0 is $p = .025$. For a two-tailed test, the region of rejection at $\alpha = .05$ consists of both tabled values of r and all values more extreme. For a one-tailed test, the region of rejection at $\alpha = .025$ consists of the tabled value of r in the predicted direction (either too small or too large) and all values more extreme.

   b. If either n_1 or n_2 is larger than 20, determine the value of z as computed from formula (4.7). Table A shows the one-tailed probability associated with the occurrence under H_0 of values as extreme as an observed z. For a two-tailed test, double the p given in that table.

If the p associated with the observed value of r is equal to or less than α, reject H_0.

Power-Efficiency

Because there are no parametric tests for the randomness of a sequence of events in a sample, the concept of power-efficiency is not meaningful in the case of the one-sample runs test.

References

For other discussions of this test, the reader is referred to Freund (1952, chap. 11), Moore and Wallis (1943), and Swed and Eisenhart (1943).

DISCUSSION

We have presented four nonparametric statistical tests of use in a one-sample design. Three of these tests are of the goodness-of-fit type, and the fourth is a test of the randomness of the sequence of events in a sample. This discussion, which compares and contrasts these tests, may aid the reader in choosing the test which will best handle the data of a given study.

In testing hypotheses about whether a sample was drawn from a population with a specified distribution, the investigator may use one of three goodness-of-fit tests: the binomial test, the χ^2 one-sample test, or the Kolmogorov-Smirnov one-sample test. His choice among these three tests should be determined by (a) the number of categories in his measurement, (b) the level of measurement used, (c) the size of the sample, and (d) the power of the statistical test.

The binomial test may be used when there are just two categories in the classification of the data. It is uniquely useful when the sample size is so small that the χ^2 test is inapplicable.

The χ^2 test should be used when the data are in discrete categories and when the expected frequencies are sufficiently large. When $k = 2$, each E_i should be 5 or larger. When $k > 2$, no more than 20 per cent of the E_i's should be smaller than 5 and none should be smaller than 1.

Both the binomial test and the χ^2 test may be used with data measured in either a nominal or an ordinal scale.

χ^2 tests are insensitive to the effects of order when $df > 1$, and thus χ^2 may not be the best test when a hypothesis takes order into account.

The Kolmogorov-Smirnov test should be used when one can assume that the variable under consideration has a continuous distribution. However, if this test is used when the population distribution, $F_0(X)$, is discontinuous, the error which occurs in the resulting probability statement is in the "safe" direction (Goodman, 1954). That is, if the tables which assume that $F_0(X)$ is continuous are used to test a hypothesis about a discontinuous variable, the test is a conservative one: if H_0 is rejected by that test we can have real confidence in that decision.

It has already been mentioned that the Kolmogorov-Smirnov test treats individual observations separately and thus does not lose information because of grouping, as the χ^2 test sometimes must. With a continuous variable, if the sample is small and therefore adjacent categories must be combined for the χ^2 test, the χ^2 test is definitely less powerful than the Kolmogorov-Smirnov test. It would seem that in all cases where it is applicable the Kolmogorov-Smirnov test is the most powerful goodness-of-fit test of those presented.

The distribution of D is not known for the case when certain parameters of the population have been estimated from the sample. However, Massey (1951a, p. 73) gives some evidence which indicates that if the Kolmogorov-Smirnov test is applied in such cases (e.g., for testing goodness-of-fit to a normal distribution with mean and standard deviation estimated from the sample), the use of Table E will lead to a conservative test. That is, if the critical value of D (as shown in Table E) is exceeded by the observed value in these circumstances, we may with considerable confidence reject H_0 (e.g., that the population is normal) and safely conclude that there is a significant difference. In cases where parameters must be estimated from the sample, the χ^2 test is easily modified for use by a reduction of the number of degrees of freedom. The Kolmogorov-Smirnov test has no such known modifications.

The one-sample runs test is concerned with the randomness of the temporal occurrence or sequence of the scores in a sample. No general statement about the efficiency of tests of randomness based on runs can be meaningful; in this case the question of efficiency has meaning only in the context of a specific problem.

CHAPTER 5

THE CASE OF TWO RELATED SAMPLES

The two-sample statistical tests are used when the researcher wishes to establish whether two treatments are different, or whether one treatment is "better" than another. The "treatment" may be any of a diverse variety of conditions: injection of a drug, training, acculturation, propaganda, separation from family, surgical alteration, changed housing conditions, intergroup integration, climate alteration, introduction of a new element into the economy, etc. In each case, the group which has undergone the treatment is compared with one which has not, or which has undergone a different treatment.

In such comparisons of two groups, sometimes significant differences are observed which are not the results of the treatment. For instance, a researcher may attempt to compare two teaching methods by having one group of students taught by one method and a different group taught by another. Now if one of the groups has abler or more motivated students, the performance of the two groups after the different learning experiences may not accurately reflect the relative effectiveness of the two teaching methods at all, because other variables are creating differences in performance.

One way to overcome the difficulty imposed by extraneous differences between groups is to use two related samples in the research. That is, one may "match" or otherwise relate the two samples studied. This matching may be achieved by using each subject as his own control, or by pairing subjects and then assigning the two members of each pair to the two conditions. When a subject "serves as his own control," he is exposed to both treatments at different times. When the pairing method is used, the effort is to select for each pair subjects who are as much alike as possible with respect to any extraneous variables which might influence the outcome of the research. In the example mentioned above, the pairing method would require that a number of pairs of students be selected, each pair composed of two students of substantially equal ability and motivation. One member of each pair, chosen from the two by random means, would be assigned to the class taught by one of the methods and his matched "partner" would be assigned to the class taught by the other method.

Wherever it is feasible, the method of using each subject as his own control (and counterbalancing the order in which the treatments are assigned) is preferable to the pairing method. The reason for this is that we are limited in our ability to match people by our ignorance of the relevant variables which determine behavior. Moreover, even when we do know what variables are important and therefore should be controlled by the pairing process, our tools for measuring these variables are rather gross or inexact and thus our pairing based on our measures may be faulty. A matching design is only as good as the experimenter's ability to determine how to match the pairs, and this ability is frequently very limited. This problem is circumvented when each subject is used as his own control; no more precise matching is possible than that achieved by identity.

The usual parametric technique for analyzing data from two related samples is to apply a t test to the difference scores. A difference score may be obtained from the two scores of the two members of each matched pair, or from the two scores of each subject under the two conditions. The t test assumes that these difference scores are normally and independently distributed in the population from which the sample was drawn, and requires that they be measured on at least an interval scale.

In a number of instances, the t test is inapplicable. The researcher may find that (a) the assumptions and requirements of the t test are unrealistic for his data, (b) he may prefer to avoid making the assumptions or testing the requirements and thus give greater generality to his conclusions, (c) his differences between matched pairs are not represented as scores but rather as "signs" (that is, he can tell which member of any pair is "greater than" the other, but cannot tell *how much* greater), or (d) his scores are simply classificatory—the two members of the matched pair can either respond in the same way or in entirely different ways which do not stand in any order or quantitative relation. In these instances, the experimenter may choose from one of the nonparametric statistical tests for two related samples which are presented in this chapter. In addition to being suitable for the cases mentioned above, these tests have the further advantage that they do not require that all pairs be drawn from the same population. Five tests are presented; the discussion at the close of the chapter indicates the special features and uses of each. This discussion may aid the reader in selecting that technique which would be most appropriate to use in a particular research.

THE McNEMAR TEST FOR THE SIGNIFICANCE OF CHANGES

Function

The McNemar test for the significance of changes is particularly applicable to those "before and after" designs in which each person is used as his own control and in which measurement is in the strength of either a nominal or ordinal scale. Thus it might be used to test the effectiveness of a particular treatment (meeting, newspaper editorial, mailed pamphlet, personal visit, etc.) on voters' preferences among various candidates. Or it might be used to test say the effects of farm-to-city moves on people's political affiliations. Notice that these are studies in which people could serve as their own controls and in which nominal measurement would be used to assess the "before to after" change.

Rationale and Method

To test the significance of any observed change by this method, one sets up a fourfold table of frequencies to represent the first and second sets of responses from the same individuals. The general features of such a table are illustrated in Table 5.1, in which $+$ and $-$ are used to signify

TABLE 5.1. FOURFOLD TABLE FOR USE IN TESTING SIGNIFICANCE OF CHANGES

After

		$-$	$+$
$+$		A	B
Before			
$-$		C	D

different responses. Notice that those cases which show changes between the first and second response appear in cells A and D. An individual is tallied in cell A if he changed from $+$ to $-$. He is tallied in cell D if he changed from $-$ to $+$. If no change is observed, he is tallied in either cell B ($+$ responses both before and after) or cell C ($-$ responses both before and after).

Since $A + D$ represents the total number of persons who changed, the expectation under the null hypothesis would be that $\frac{1}{2}(A + D)$ cases changed in one direction and $\frac{1}{2}(A + D)$ cases changed in the other. In other words, $\frac{1}{2}(A + D)$ is the expected frequency under H_0 in both cell A and cell D.

It will be remembered from Chap. 4 that

$$\chi^2 = \sum_{i=1}^{k} \frac{(O_i - E_i)^2}{E_i} \tag{4.5}$$

where O_i = observed number of cases in ith category

E_i = expected number of cases under H_0 in ith category

In the McNemar test for the significance of changes, we are interested only in cells A and D. Therefore, if A = the observed number of cases in cell A, D = the observed number of cases in cell D, and $\frac{1}{2}(A + D)$ = the expected number of cases in both cell A and cell D, then

$$\chi^2 = \sum_{A,D} \frac{(O - E)^2}{E}$$

$$= \frac{\left(A - \dfrac{A + D}{2}\right)^2}{\dfrac{A + D}{2}} + \frac{\left(D - \dfrac{A + D}{2}\right)^2}{\dfrac{A + D}{2}}$$

Expanding and collecting terms, we have

$$\chi^2 = \frac{(A - D)^2}{A + D} \quad \text{with } df = 1 \tag{5.1}$$

That is, the sampling distribution under H_0 of χ^2 as yielded by formula (5.1) is distributed approximately as chi square[1] with $df = 1$.

Correction for continuity. The approximation by the chi-square distribution of the sampling distribution of formula (5.1) becomes an excellent one if a correction for continuity is performed. The correction is necessary because a continuous distribution (chi square) is used to approximate a discrete distribution. When all expected frequencies are small, that approximation may be a poor one. The correction for continuity (Yates, 1934) is an attempt to remove this source of error.

With the correction for continuity included

$$\chi^2 = \frac{(|A - D| - 1)^2}{A + D} \quad \text{with } df = 1 \tag{5.2}$$

This expression directs one to subtract 1 from the absolute value of the difference between A and D (that is, from that difference irrespective of sign) before squaring. The significance of any observed value of χ^2, as computed from formula (5.2), is determined by reference to Table C of the Appendix which gives various critical values of chi square for df's from 1 to 30. That is, if the observed value of χ^2 is equal to or greater than that shown in Table C for a particular significance level at $df = 1$, the implication is that a "significant" effect was demonstrated in the "before" and "after" responses.

The example which follows illustrates the application of this test.

[1] See footnote on page 43.

Example

Suppose a child psychologist is interested in children's initiation of social contacts. He has observed that children who are new in a nursery school usually initiate interpersonal contacts with adults rather than with other children. He predicts that with increasing familiarity and experience, children will increasingly initiate social contacts with other children rather than with adults. To test this hypothesis, he observes 25 new children on each child's first day at nursery school, and he categorizes their first initiation of social contact according to whether it was directed to an adult or to another child. He then observes each of the 25 children after each has attended nursery school for a month, making the same categorization. Thus his data are cast in the form shown in Table 5.2. His test of the hypothesis follows; this discussion reports artificial data.

TABLE 5.2. FORM OF FOURFOLD TABLE TO SHOW CHANGES IN
CHILDREN'S OBJECTS OF INITIATION

| | | Object of initiation on thirtieth day | |
		Child	Adult
Object of initiation on first day	Adult	A	B
	Child	C	D

i. *Null Hypothesis.* H_0: for those children who change, the probability that any child will change his object of initiation from adult to child (that is, P_A) is equal to the probability that he will change his object of initiation from child to adult (that is, P_D) is equal to one-half. That is, $P_A = P_D = \frac{1}{2}$.* H_1: $P_A > P_D$.

ii. *Statistical Test.* The McNemar test for the significance of changes is chosen because the study uses two related samples, is of the before-and-after type, and uses nominal (classificatory) measurement.

iii. *Significance Level.* Let $\alpha = .05$. $N = 25$, the number of children observed on the first and thirtieth day at school of each.

iv. *Sampling Distribution.* Table C gives critical values of chi-square for various levels of significance. The sampling distribution of χ^2 as computed by formula (5.2) is very closely approximated by the chi-square distribution with $df = 1$.

v. *Rejection Region.* Since H_1 specifies the direction of the pre-

---

* This statement of H_0 suggests a straightforward application of the binomial test (pages 36 to 42). The relation of the McNemar test to the binomial test is shown in the discussion of small expected frequencies (below).

dicted difference, the region of rejection is one-tailed. The region of rejection consists of all values of χ^2 (computed from data in which $A > D$) which are so large that they have a one-tailed probability associated with their occurrence under H_0 of .05 or less.

vi. *Decision.* The artificial data of this study are shown in Table 5.3. It shows that $A = 14 =$ the number of children whose

TABLE 5.3. CHILDREN'S OBJECTS OF INITIATION ON FIRST AND THIRTIETH DAYS IN NURSERY SCHOOL
(Artificial data)

| | | Object of initiation on thirtieth day | |
		Child	Adult
Object of initiation on first day	Adult	14	4
	Child	3	4

objects changed from adult to child, and $D = 4 =$ the number of children whose objects changed from child to adult. $B = 4$ and $C = 3$ represent those children whose objects were in the same category on both occasions. We are interested in the children who showed change: those represented in cells A and D.

For these data,

$$\chi^2 = \frac{(|A - D| - 1)^2}{A + D} \tag{5.2}$$

$$= \frac{(|14 - 4| - 1)^2}{14 + 4}$$

$$= \frac{9^2}{18}$$

$$= 4.5$$

Reference to Table C reveals that when $\chi^2 \geq 4.5$ and $df = 1$, the probability of occurrence under H_0 is $p < \frac{1}{2}(.05)$ which is $p < .025$. (The probability value given in Table C is halved because a one-tailed test is called for and the table gives two-tailed values.)

Inasmuch as the probability under H_0 associated with the occurrence we observed is $p < .025$ and is less than $\alpha = .05$, the observed value of χ^2 is in the region of rejection and thus our decision is to reject H_0 in favor of H_1. With these artificial data we conclude that children show a significant tendency to change their objects of initiation from adults to children after 30 days of nursery school experience.

Small expected frequencies. If the expected frequency, that is, $\frac{1}{2}(A + D)$, is very small (less than 5), the binomial test (Chap. 4) should be used rather than the McNemar test.   For the binomial test, $N = A + D$, and $x =$ the smaller of the two observed frequencies, either A or D.

Notice that we could have tested the data in Table 5.3 with the binomial test.   The null hypothesis would be that the sample of $N = A + D$ cases came from a binomial population where $P = Q = \frac{1}{2}$.   For the above data, $N = 18$ and $x = 4$, the smaller of the two frequencies observed. Table D of the Appendix shows the probability under H_0 associated with such a small value is $p = .015$ which is essentially the same p yielded by the McNemar test.   The difference between the two p's is due mainly to the fact that the chi-square table does not include all values between $p = .05$ and $p = .01$.

Summary of procedure.   These are the steps in the computation of the McNemar test:

1. Cast the observed frequencies in a fourfold table of the form illustrated in Table 5.1.

2. Determine the expected frequencies in cells A and D.

$$E = \tfrac{1}{2}(A + D)$$

If the expected frequencies are less than 5, use the binomial test rather than the McNemar test.

3. If the expected frequencies are 5 or larger, compute the value of χ^2 using formula (5.2).

4. Determine the probability under H_0 associated with a value as large as the observed value of χ^2 by referring to Table C.   If a one-tailed test is called for, halve the probability shown in that table.   If the p shown by Table C for the observed value of χ^2 with $df = 1$ is equal to or less than α, reject H_0 in favor of H_1.

Power-Efficiency

When the McNemar test is used with nominal measures, the concept of power-efficiency is meaningless inasmuch as there is no alternative with which to compare the test.   However, when the measurement and other aspects of the data are such that it is possible to apply the parametric t test, the McNemar test, like the binomial test, has power-efficiency of about 95 per cent for $A + D = 6$, and the power-efficiency declines as the size of $A + D$ increases to an eventual (asymptotic) efficiency of about 63 per cent.

References

Discussions of this test are presented by Bowker (1948) and McNemar (1947; 1955, pp. 228–231).

THE SIGN TEST

Function

The sign test gets its name from the fact that it uses plus and minus signs rather than quantitative measures as its data. It is particularly useful for research in which quantitative measurement is impossible or infeasible, but in which it is possible to rank with respect to each other the two members of each pair.

The sign test is applicable to the case of two related samples when the experimenter wishes to establish that two conditions are different. The only assumption underlying this test is that the variable under consideration has a continuous distribution. The test does not make any assumptions about the form of the distribution of differences, nor does it assume that all subjects are drawn from the same population. The different pairs may be from different populations with respect to age, sex, intelligence, etc.; the only requirement is that within each pair the experimenter has achieved matching with respect to the relevant extraneous variables. As was noted before, one way of accomplishing this is to use each subject as his own control.

Method

The null hypothesis tested by the sign test is that

$$p(X_A > X_B) = p(X_A < X_B) = \tfrac{1}{2}$$

where X_A is the judgment or score under one of the conditions (or after the treatment) and X_B is the judgment or score under the other condition (or before the treatment). That is, X_A and X_B are the two "scores" for a matched pair. Another way of stating H_0 is: the median difference is zero.

In applying the sign test, we focus on the direction of the differences between every X_{Ai} and X_{Bi}, noting whether the sign of the difference is plus or minus. Under H_0, we would expect the number of pairs which have $X_A > X_B$ to equal the number of pairs which have $X_A < X_B$. That is, if the null hypothesis were true we would expect about half of the differences to be negative and half to be positive. H_0 is rejected if too few differences of one sign occur.

Small samples. The probability associated with the occurrence of a particular number of $+$'s and $-$'s can be determined by reference to the binomial distribution with $P = Q = \tfrac{1}{2}$, where $N =$ the number of pairs. If a matched pair shows no difference (i.e., the difference, being zero, has no sign) it is dropped from the analysis and N is thereby reduced. Table D of the Appendix gives the probabilities associated with the occurrence under H_0 of values as small as x for $N \leq 25$. To use this table, let x = the number of fewer signs.

For example, suppose 20 pairs are observed. Sixteen show differences in one direction ($+$) and the other four show differences in the other ($-$). Here $N = 20$ and $x = 4$. Reference to Table D reveals that the probability of this distribution of $+$'s and $-$'s or an even more extreme one under H_0 is $p = .006$ (one-tailed).

The sign test may be either one-tailed or two-tailed. In a one-tailed test, the advance prediction states which sign, $+$ or $-$, will occur more frequently. In a two-tailed test, the prediction is simply that the frequencies with which the two signs occur will be significantly different. For a two-tailed test, double the values of p shown in Table D.

Example for Small Samples

In a study of the effects of father-absence upon the development of children, 17 married couples who had been separated by war and whose first child was born during the father's absence were interviewed, husbands and wives separately. Each was asked to discuss various topics concerning the child whose first year had been spent in a fatherless home. Each parent was asked to discuss the father's disciplinary relations with the child in the years after his return from war. These statements were extracted from the recorded interviews, and a psychologist who knew each family was asked to rate the statements on the degree of insight which each parent showed in discussing paternal discipline.[1] The prediction was that the mother, because of her longer and closer association with the child and because of a variety of other circumstances typically associated with father-separation because of war, would have greater insight into her husband's disciplinary relations with their child than he would have.

i. *Null Hypothesis.* H_0: the median of the differences is zero. That is, there are as many husbands whose insight into their own disciplinary relations with their children is greater than their wives' as there are wives whose insight into paternal discipline is greater than their husbands'. H_1: the median of the differences is positive.

ii. *Statistical Test.* The rating scale used in this study constituted at best a partially ordered scale. The information contained in the ratings is preserved if the difference between each couple's two ratings is expressed by a sign. Each married couple in this study constitutes a matched pair; they are matched in the sense that each

---

[1] Engvall, Alberta. 1954. Comparison of mother and father attitudes toward war-separated children. In Lois M. Stolz et al., *Father relations of war-born children.* Stanford, Calif.: Stanford Univer. Press. Pp. 149–180.

discussed the same child and the same family situation in the material rated. The sign test is appropriate for measures of the strength indicated, and of course is appropriate for a case of two related samples.

iii. *Significance Level.* Let $\alpha = .05$. $N = 17$, the number of war-separated couples. (N may be reduced if ties occur.)

iv. *Sampling Distribution.* The associated probability of occurrence of values as small as x is given by the binomial distribution for $P = Q = \frac{1}{2}$. The associated probabilities are given in Table D.

v. *Rejection Region.* Since H_1 predicts the direction of the differences, the region of rejection is one-tailed. It consists of all values of x (where $x =$ the number of minuses, since the prediction is that pluses will predominate and $x =$ the number of fewer signs) whose one-tailed associated probability of occurrence under H_0 is equal to or less than $\alpha = .05$.

vi. *Decision.* The statements of each parent were rated on a five-point rating scale. On this scale, a rating of 1 represents high insight.

TABLE 5.4. WAR-SEPARATED PARENTS' INSIGHT INTO PATERNAL DISCIPLINE

Couple (pseudonym)	Rating on insight* into paternal discipline		Direction of difference	Sign.
	F	M		
Mr. and Mrs. Arnold	4	2	$X_F > X_M$	+
Mr. and Mrs. Brown	4	3	$X_F > X_M$	+
Mr. and Mrs. Burgman	5	3	$X_F > X_M$	+
Mr. and Mrs. Ford	5	3	$X_F > X_M$	+
Mr. and Mrs. Harlow	3	3	$X_F = X_M$	0
Mr. and Mrs. Holman	2	3	$X_F < X_M$	−
Mr. and Mrs. Irwin	5	3	$X_F > X_M$	+
Mr. and Mrs. Marston	3	3	$X_F = X_M$	0
Mr. and Mrs. Mathews	1	2	$X_F < X_M$	−
Mr. and Mrs. Moore	5	3	$X_F > X_M$	+
Mr. and Mrs. Osborne	5	2	$X_F > X_M$	+
Mr. and Mrs. Snyder	5	2	$X_F > X_M$	+
Mr. and Mrs. Soule	4	5	$X_F < X_M$	−
Mr. and Mrs. Statler	5	2	$X_F > X_M$	+
Mr. and Mrs. Wagner	5	5	$X_F = X_M$	0
Mr. and Mrs. Wolf	5	3	$X_F > X_M$	+
Mr. and Mrs. Wycoff	5	1	$X_F > X_M$	+

* A rating of 1 represents great insight; a rating of 5 represents little or no insight.

Table 5.4 shows the ratings assigned to each mother (M) and father (F) among the 17 war-separated couples. The signs of the differences between each couple are shown in the final column. Observe that 3 couples (the Holmans, Mathewses, and Soules) showed differences in the opposite direction from that predicted, i.e., in each case $X_F < X_M$, and thus each of these 3 received a minus. For 3 couples (the Harlows, Marstons, and Wagners), there was no difference between the two ratings, that is, $X_F = X_M$, and thus these couples received no sign. The remaining 11 couples showed differences in the predicted direction.

For the data in Table 5.4, x = the number of fewer signs = 3, and N = the number of matched pairs who showed differences = 14. Table D shows that for $N = 14$, an $x \le 3$ has a one-tailed probability of occurrence under H_0 of $p = .029$. This value is in the region of rejection for $\alpha = .05$; thus our decision is to reject H_0 in favor of H_1. We conclude that war-separated wives show greater insight into their husbands' disciplinary relations with their war-born children than do the husbands themselves.

Ties. For the sign test, a "tie" occurs when it is not possible to discriminate between a matched pair on the variable under study, or when the two scores earned by any pair are equal. In the case of the war-separated couples, three ties occurred: the psychologist rated three couples as having equal insight into paternal discipline.

All tied cases are dropped from the analysis for the sign test, and the N is correspondingly reduced. Thus N = the number of matched pairs whose difference score *has a sign*. In the example, 14 of the 17 couples had difference scores with a sign, so for that case $N = 14$.

Relation to the binomial expansion. In the study just discussed, we should expect under H_0 that the frequency of pluses and minuses would be the same as the frequency of heads and tails in a toss of 14 unbiased coins. (More exactly, the analogy is to the toss of 17 unbiased coins, 3 of which rolled out of sight and thus could not be included in the analysis.) The probability of getting as extreme an occurrence as 3 heads and 11 tails in a toss of 14 coins is given by the binomial distribution as

$$\sum_{x=0}^{3} \binom{N}{x} P^x Q^{N-x}$$

where N = total number of coins
x = observed number of heads
$$\binom{N}{x} = \frac{N!}{x! \, (N-x)!}$$

In the case of 3 or fewer heads when 14 coins are tossed, this is

$$p = \frac{\binom{14}{0} + \binom{14}{1} + \binom{14}{2} + \binom{14}{3}}{2^{14}}$$

$$= \frac{1 + 14 + 91 + 364}{16,284}$$

$$= .029$$

The probability value found by this method is of course identical to that found by the method used in the example: $p = .029$.

Large samples. If N is larger than 25, the normal approximation to the binomial distribution can be used. This distribution has

$$\text{Mean} = \mu_x = NP = \tfrac{1}{2}N$$

and $\text{Standard deviation} = \sigma_x = \sqrt{NPQ} = \tfrac{1}{2}\sqrt{N}$

That is, the value of z is given by

$$z = \frac{x - \mu_x}{\sigma_x} = \frac{x - \tfrac{1}{2}N}{\tfrac{1}{2}\sqrt{N}} \tag{5.3}$$

This expression is approximately normally distributed with zero mean and unit variance.

The approximation becomes an excellent one when a *correction for continuity* is employed. The correction is effected by reducing the difference between the observed number of pluses (or minuses) and the expected number, i.e., the mean under H_0, by .5. (See pages 40 to 41 for a more complete discussion of this point.) That is, with the correction for continuity

$$z = \frac{(x \pm .5) - \tfrac{1}{2}N}{\tfrac{1}{2}\sqrt{N}} \tag{5.4}$$

where $x + .5$ is used when $x < \tfrac{1}{2}N$, and $x - .5$ is used when $x > \tfrac{1}{2}N$. The value of z obtained by the application of formula (5.4) may be considered to be normally distributed with zero mean and unit variance. Therefore the significance of an obtained z may be determined by reference to Table A in the Appendix. That is, Table A gives the one-tailed probability associated with the occurrence under H_0 of values as extreme as an observed z. (If a two-tailed test is required, the p yielded by Table A should be doubled.)

Example for Large Samples

Suppose an experimenter were interested in determining whether a certain film about juvenile delinquency would change the opinions of the members of a particular community about how severely juvenile delinquents should be punished. He draws a random sample of 100 adults from the community, and conducts a "before and after" study, having each subject serve as his own control. He asks each subject to take a position on whether *more* or *less* punitive action against juvenile delinquents should be taken than is taken at present. He then shows the film to the 100 adults, after which he repeats the question.

i. *Null Hypothesis.* H_0: the film has no systematic effect. That is, of those whose opinions change after seeing the film, just as many change from *more* to *less* as change from *less* to *more*, and any difference observed is of a magnitude which might be expected in a random sample from a population on which the film would have no systematic effect. H_1: the film has a systematic effect.

ii. *Statistical Test.* The sign test is chosen for this study of two related groups because the study uses ordinal measures within matched pairs, and therefore the differences may appropriately be represented by plus and minus signs.

iii. *Significance Level.* Let $\alpha = .01$. $N =$ the number of subjects (out of 100) who show an opinion change in either direction.

iv. *Sampling Distribution.* Under H_0, z as computed from formula (5.4) is approximately normally distributed for $N > 25$. Table A gives the probability associated with the occurrence of values as extreme as an obtained z.

v. *Rejection Region.* Since H_1 does not state the direction of the predicted differences, the region of rejection is two-tailed. It consists of all values of z which are so extreme that their associated probability of occurrence under H_0 is equal to or less than $\alpha = .01$.

vi. *Decision.* The results of this hypothetical study of the effects of a film upon opinion are shown in Table 5.5.

TABLE 5.5. ADULT OPINIONS CONCERNING WHAT SEVERITY OF PUNISHMENT IS DESIRABLE FOR JUVENILE DELINQUENCY
(Artificial data)

		Amount of punishment favored after film	
		Less	More
Amount of punishment favored before film	More	59	7
	Less	8	26

Did the film have any effect? The data show that there were 15 adults (8 + 7) who were unaffected and 85 who were. The hypothesis of the study applies only to those 85. If the film had no systematic effect, we would expect about half of those whose opinions changed from before to after to have changed from *more* to *less*, and about half to have changed from *less* to *more*. That is, we would expect about 42.5 subjects to show each of the two kinds of change. Now we observe that 59 changed from *more* to *less*, while 26 changed from *less* to *more*. We may determine the associated probability under H_0 of such an extreme split by using formula (5.4). For these data, $x > \frac{1}{2}N$, that is, $59 > 42.5$.

$$z = \frac{(x \pm .5) - \frac{1}{2}N}{\frac{1}{2}\sqrt{N}} \tag{5.4}$$

$$= \frac{(59 - .5) - \frac{1}{2}(85)}{\frac{1}{2}\sqrt{85}}$$

$$= 3.47$$

Reference to Table A reveals that the probability under H_0 of $z \geq 3.47$ is $p = 2(.0003) = .0006$. (The p shown in the table is doubled because the tabled values are for a one-tailed test, whereas the region of rejection in this case is two-tailed.) Inasmuch as $p = .0006$ is smaller than $\alpha = .01$, the decision is to reject the null hypothesis in favor of the alternative hypothesis. We conclude from these fictitious data that the film had a significant systematic effect on adults' opinions regarding the severity of punishment which is desirable for juvenile delinquents.

This example was included not only because it demonstrates a useful application of the sign test, but also because data of this sort are often analyzed incorrectly. It is not too uncommon for researchers to analyze such data by using the row and column totals as if they represented independent samples. This is not the case; the row and column totals are separate but not independent representations of the same data.

This example could also have been analyzed by the McNemar test for the significance of changes (discussed on pages 63 to 67). With the data shown in Table 5.5,

$$\chi^2 = \frac{(|A - D| - 1)^2}{A + D} \tag{5.2}$$

$$= \frac{(|59 - 26| - 1)^2}{59 + 26}$$

$$= 12.05$$

Table C shows that $\chi^2 \geq 12.05$ with $df = 1$ has a probability of occurrence under H_0 of $p < .001$. This finding is not in conflict with that yielded by the sign test. The difference between the two findings is due to the limitations of the chi-square table used.

Summary of procedure. These are the steps in the use of the sign test:

1. Determine the sign of the difference between the two members of each pair.

2. By counting, determine the value of N = the numbers of pairs whose differences show a sign.

3. The method for determining the probability associated with the occurrence under H_0 of a value as extreme as the observed value of x depends on the size of N:

   $a.$ If N is 25 or smaller, Table D shows the one-tailed p associated with a value as small as the observed value of x = the number of fewer signs. For a two-tailed test, double the value of p shown in Table D.

   $b.$ If N is larger than 25, compute the value of z, using formula (5.4). Table A gives one-tailed p's associated with values as extreme as various values of z. For a two-tailed test, double the value of p shown in Table A.

If the p yielded by the test is equal to or less than α, reject H_0.

Power-Efficiency

The power-efficiency of the sign test is about 95 per cent for $N = 6$, but it declines as the size of the sample increases to an eventual (asymptotic) efficiency of 63 per cent. For discussions of the power-efficiency of the sign test for large samples, see Mood (1954) and Walsh (1946).

References

For other discussions of the sign test, the reader is directed to Dixon and Massey (1951, chap. 17), Dixon and Mood (1946), McNemar (1955, pp. 357–358), Moses (1952a), and Walsh, (1946).

THE WILCOXON MATCHED-PAIRS SIGNED-RANKS TEST

Function

The test we have just discussed, the sign test, utilizes information simply about the *direction* of the differences within pairs. If the relative *magnitude* as well as the direction of the differences is considered, a more powerful test can be made. The Wilcoxon matched-pairs signed-ranks test does just that: it gives more weight to a pair which shows a large difference between the two conditions than to a pair which shows a small difference.

The Wilcoxon test is a most useful test for the behavioral scientist. With behavioral data, it is not uncommon that the researcher can (*a*) tell which member of a pair is "greater than" which, i.e., tell the sign of the difference between any pair, and (*b*) rank the differences in order of absolute size. That is, he can make the judgment of "greater than" between any pair's two performances, and also can make that judgment between any two difference scores arising from any two pairs. With such information,[1] the experimenter may use the Wilcoxon test.

Rationale and Method

Let d_i = the difference score for any matched pair, representing the difference between the pair's scores under the two treatments. Each pair has one d_i. To use the Wilcoxon test, rank all the d_i's without regard to sign: give the rank of 1 to the smallest d_i, the rank of 2 to the next smallest, etc. When one ranks scores without respect to sign, a d_i of -1 is given a lower rank than a d_i of either -2 or $+2$.

Then to each *rank* affix the sign of the difference. That is, indicate which ranks arose from negative d_i's and which ranks arose from positive d_i's.

Now if treatments A and B are equivalent, that is, if H_0 is true, we should expect to find some of the larger d_i's favoring treatment A and some favoring treatment B. That is, some of the larger ranks would come from positive d_i's while others would come from negative d_i's. Thus, if we summed the ranks having a plus sign and summed the ranks having a minus sign, we would expect the two sums to be about equal under H_0. But if the sum of the positive ranks is very much different from the sum of the negative ranks, we would infer that treatment A differs from treatment B, and thus we would reject H_0. That is, we reject H_0 if either the sum of the ranks for the negative d_i's *or* the sum of the ranks for the positive d_i's is too small.

Ties. Occasionally the two scores of any pair are equal. That is, no difference between the two treatments is observed for that pair, so that $d = 0$. Such pairs are dropped from the analysis. This is the same practice that we follow with the sign test. N = the number of matched pairs minus the number of pairs whose $d = 0$.

Another sort of tie can occur. Two or more d's can be of the same size. We assign such tied cases the same rank. The rank assigned is the *average of the ranks* which would have been assigned if the d's had differed slightly. Thus three pairs might yield d's of -1, -1, and $+1$. Each pair would be assigned the rank of 2, for $\dfrac{1 + 2 + 3}{3} = 2$. Then the next d in order would receive the rank of 4, because ranks 1, 2, and

---

[1] To require that the researcher have ordinal information not only within pairs but also concerning the differences between pairs seems to be tantamount to requiring measurement in the strength of an *ordered metric* scale. In strength, an ordered metric scale lies between an ordinal scale and an interval scale. For a discussion of ordered metric scaling, see Coombs (1950) and Siegel (1956).

3 have already been used. If two pairs had yielded d's of 1, both would receive the rank of 1.5, and the next largest d would receive the rank of 3. The practice of giving tied observations the average of the ranks they would otherwise have gotten has a negligible effect on T, the statistic on which the Wilcoxon test is based.

For applications of these principles for the handling of ties, see the example for large samples, later in this section.

Small samples. Let T = the smaller sum of like-signed ranks. That is, T is either the sum of the positive ranks or the sum of the negative ranks, whichever sum is smaller. Table G of the Appendix gives various values of T and their associated levels of significance. That is, if an observed T is equal to or less than the value given in the body of Table G under a particular significance level for the observed value of N, the null hypothesis may then be rejected at that level of significance.

Table G is adapted for use with both one-tailed and two-tailed tests. A one-tailed test may be used if in advance of examining the data the experimenter predicts the sign of the smaller sum of ranks. That is, as is the case with all one-tailed tests, he must predict in advance the direction of the differences.

For example, if $T = 3$ were the sum of the negative ranks when $N = 9$, one could reject H_0 at the $\alpha = .02$ level if H_1 had been that the two groups would differ, and one could reject H_0 at the $\alpha = .01$ level if H_1 had been that the sum of negative ranks would be the smaller sum.

Example for Small Samples

Suppose a child psychologist wished to test whether nursery school attendance has any effect on children's social perceptiveness. He scores social perceptiveness by rating children's responses to a group of pictures which depict a variety of social situations, asking a standard group of questions about each picture. By this device he obtains a score between 0 and 100 for each child.

Although the experimenter is confident that a higher score represents higher social perceptiveness than a lower score, he is not sure that the scores are sufficiently exact to be treated numerically. That is, he is not willing to say that a child whose score is 60 is *twice* as socially perceptive as a child whose score is 30, nor is he willing to say that the difference between scores of 60 and 40 is exactly twice as large as the difference between scores of 40 and 30. However, he is confident that the difference between a score of, say, 60 and one of 40 is greater than the difference between a score of 40 and one of 30. That is, he cannot assert that the differences are numerically exact, but he does maintain that they are sufficiently meaningful that they may appropriately be ranked in order of absolute size.

To test the effect of nursery school attendance on children's social perceptiveness scores, he obtains 8 pairs of identical twins to serve as subjects. At random, 1 twin from each pair is assigned to attend nursery school for a term. The other twin in each pair is to remain out of school. At the end of the term, the 16 children are each given the test of social perceptiveness.

i. *Null Hypothesis.* H_0: the social perceptiveness of "home" and "nursery school" children does not differ. In terms of the Wilcoxon test, the sum of the positive ranks = the sum of the negative ranks. H_1: the social perceptiveness of the two groups of children differs, i.e., the sum of the positive ranks $\neq$ the sum of the negative ranks.

ii. *Statistical Test.* The Wilcoxon matched-pairs signed-ranks test is chosen because the study employs two related samples and it yields difference scores which may be ranked in order of absolute magnitude.

iii. *Significance Level.* Let $\alpha = .05$. N = the number of pairs (8) minus any pairs whose d is zero.

iv. *Sampling Distribution.* Table G gives critical values from the sampling distribution of T, for $N \leq 25$.

v. *Rejection Region.* Since the direction of the difference is not predicted, a two-tailed region of rejection is appropriate. The region of rejection consists of all values of T which are so small that the probability associated with their occurrence under H_0 is equal to or less than $\alpha = .05$ for a two-tailed test.

vi. *Decision.* In this fictitious study, the 8 pairs of "home" and "nursery school" children are given the test in social perceptiveness after the latter have been in school for one term. Their scores are given in Table 5.6. The table shows that only 2 pairs of twins, c and g, showed differences in the direction of greater social perceptiveness in the "home" twin. And these difference scores are among the smallest: their ranks are 1 and 3.

The smaller of the sums of the like-signed ranks = $1 + 3 = 4 = T$. Table G shows that for $N = 8$, a T of 4 allows us to reject the null hypothesis at $\alpha = .05$ for a two-tailed test. Therefore we reject H_0 in favor of H_1 in this fictitious study, concluding that nursery school experience does affect the social perceptiveness of children.

It is worth noting that the data in Table 5.6 are amenable to treatment by the sign test (pages 68 to 75), a less powerful test. For that test, $x = 2$ and $N = 8$. Table D gives the probability associated with such an occurrence under H_0 as $p = 2(.145) = .290$ for a two-tailed test. With the sign test, therefore, our decision would be to *accept* H_0 when $\alpha = .05$,

TABLE 5.6. SOCIAL PERCEPTIVENESS SCORES OF "NURSERY SCHOOL"
AND "HOME" CHILDREN
(Artificial data)

Pair	Social perceptiveness score of twin in nursery school	Social perceptiveness score of twin at home	d	Rank of d	Rank with less frequent sign
a	82	63	19	7	
b	69	42	27	8	
c	73	74	−1	−1	1
d	43	37	6	4	
e	58	51	7	5	
f	56	43	13	6	
g	76	80	−4	−3	3
h	65	82	3	2	
					$T = 4$

whereas the Wilcoxon test enabled us to *reject* H_0 at that level. This difference is not surprising, for the Wilcoxon test utilizes more of the information in the data. Notice that the Wilcoxon test takes into consideration the fact that the 2 minus d's are among the smallest d's observed, whereas the sign test is unaffected by the relative magnitude of the d_i's.

Large samples. When N is larger than 25, Table G cannot be used. However, it can be shown that in such cases the sum of the ranks, T, is practically normally distributed, with

$$\text{Mean} = \mu_T = \frac{N(N + 1)}{4}$$

and   Standard deviation $= \sigma_T = \sqrt{\dfrac{N(N + 1)(2N + 1)}{24}}$

Therefore $z = \dfrac{T - \mu_T}{\sigma_T} = \dfrac{T - \dfrac{N(N + 1)}{4}}{\sqrt{\dfrac{N(N + 1)(2N + 1)}{24}}}$ (5.5)

is approximately normally distributed with zero mean and unit variance. Thus Table A of the Appendix gives the probabilities associated with the occurrence under H_0 of various values as extreme as an observed z computed from formula (5.5).

To show what an excellent approximation this is, even for small samples, we shall treat the data given in Table 5.6, where $N = 8$, by this large-sample approximation. In that case, $T = 4$. Inserting the values in formula (5.5), we have

$$z = \frac{4 - \frac{(8)(9)}{4}}{\sqrt{\frac{(8)(9)(17)}{24}}}$$

$$= -1.96$$

Reference to Table A reveals that the probability associated with the occurrence under H_0 of a z as extreme as -1.96 is $p = 2(.025) = .05$, for a two-tailed test. This is the same p we found by using Table G for the same data.

Example for Large Samples

Inmates in a federal prison served as subjects in a decision-making study.[1] First the prisoners' utility (subjective value) for cigarettes was measured individually, cigarettes being negotiable in prison society. Using each subject's utility function, the experimenter then attempted to predict the decisions the man would make in a game in which he repeatedly had to choose between two different (varying) gambles, and in which cigarettes might be won or lost.

The first hypothesis tested was that the experimenter could predict the subjects' decisions by means of their utility functions better than he could by assuming that their utility for cigarettes was equal to the cigarettes' objective value and therefore predicting the "rational" choice in terms of objective value. This hypothesis was confirmed.

However, as was expected, some responses were not predicted successfully by this hypothesis of maximization of expected utility. Anticipating this outcome, the experimenter had hypothesized that such errors in prediction would be due to the indifference of the subjects between the two gambles offered. That is, a prisoner might find two gambles either equally attractive or equally unattractive, and therefore be indifferent in the choice between them. Such choices would be difficult to predict. But in such choices, it was reasoned that the subject might vacillate considerably before stating a decision. That is, the latency time between the offer of the gamble and his statement of a decision would be high. The second hypothesis, then, was that the latency times for those choices which would not be predicted successfully by maximization of expected utility would be longer than the latency times for those choices which would be successfully predicted.

i. *Null Hypothesis.* H_0: there is no difference between the latency times of incorrectly predicted and correctly predicted decisions. H_1:

---

[1] Hurst, P. M., and Siegel, S. 1956. Prediction of decisions from a higher ordered metric scale of utility. *J. exp. Psychol.*, **52**, 138–144.

the latency times of incorrectly predicted decisions are longer than the latency times of correctly predicted decisions.

ii. *Statistical Test.* The Wilcoxon matched-pairs signed-ranks test is selected because the data are difference scores from two related samples (correctly predicted choices and incorrectly predicted choices made by the same prisoners), where each subject is used as his own control.

iii. *Significance Level.* Let $\alpha = .01$. $N = 30$ = the number of prisoners who served as subjects. (This N will be reduced if any prisoner's d is zero.)

iv. *Sampling Distribution.* Under H_0, the values of z as computed from formula (5.5) are normally distributed with zero mean and unit variance. Thus Table A gives the probability associated with the occurrence under H_0 of values as extreme as an obtained z.

v. *Rejection Region.* Since the direction of the difference is predicted, the region of rejection is one-tailed. If the difference is in the predicted direction, T, the smaller of the sums of the like-signed ranks, will be the sum of the ranks of those prisoners whose d's are in the opposite direction from that predicted. The region of rejection consists of all z's (obtained from data with such T's) which are so extreme that the probability associated with their occurrence under H_0 is equal to or less than $\alpha = .01$.

vi. *Decision.* A difference score (d) was obtained for each subject by subtracting his median time in coming to correctly predicted decisions from his median time in coming to incorrectly predicted decisions. Table 5.7 gives these values of d for the 30 prisoners, and gives the other information necessary for computing the Wilcoxon test. A minus d indicates that the prisoner's median time in coming to correctly predicted decisions was *longer* than his median time in coming to incorrectly predicted decisions.

For the data in Table 5.7, $T = 53.0$, the smaller of the sums of the like-signed ranks. We apply formula (5.5):

$$z = \frac{T - \dfrac{N(N + 1)}{4}}{\sqrt{\dfrac{N(N + 1)(2N + 1)}{24}}} \tag{5.5}$$

$$= \frac{53 - \dfrac{(26)(27)}{4}}{\sqrt{\dfrac{(26)(27)(53)}{24}}}$$

$$= -3.11$$

TABLE 5.7. DIFFERENCE IN MEDIAN TIME BETWEEN PRISONERS'
CORRECTLY AND INCORRECTLY PREDICTED DECISIONS

Prisoner	d	Rank of d	Rank with less frequent sign
1	−2	−11.5	11.5
2	0		
3	0		
4	1	4.5	
5	0		
6	0		
7	4	20.0	
8	4	20.0	
9	1	4.5	
10	1	4.5	
11	5	23.0	
12	3	16.5	
13	5	23.0	
14	3	16.5	
15	−1	−4.5	4.5
16	1	4.5	
17	−1	−4.5	4.5
18	5	23.0	
19	8	25.5	
20	2	11.5	
21	2	11.5	
22	2	11.5	
23	−3	−16.5	16.5
24	−2	−11.5	11.5
25	1	4.5	
26	4	20.0	
27	8	25.5	
28	2	11.5	
29	3	16.5	
30	−1	−4.5	4.5
			$T = 53.0$

Notice that we have $N = 26$, for 4 of the prisoners' median times
were the same for both correctly and incorrectly predicted decisions
and thus their d's were 0. Notice also that our T is the sum of the
ranks of those prisoners whose d's are in the opposite direction from
predicted, and therefore we are justified in proceeding with a one-
tailed test. Table A shows that z as extreme as −3.11 has a one-
tailed probability associated with its occurrence under H_0 of $p = .0009$.
Inasmuch as this p is less than $\alpha = .01$ and thus the value of z is in
the region of rejection, our decision is to reject H_0 in favor of H_1.
We conclude that the prisoners' latency times for incorrectly pre-

dicted decisions were significantly longer than their latency times for correctly predicted decisions. This conclusion lends some support to the idea that the incorrectly predicted decisions concerned gambles which were equal, or approximately equal, in expected utility to the subjects.

Summary of procedure. These are the steps in the use of the Wilcoxon matched-pairs signed-ranks test:

1. For each matched pair, determine the signed difference (d_i) between the two scores.
2. Rank these d_i's without respect to sign. With tied d's, assign the average of the tied ranks.
3. Affix to each rank the sign $(+$ or $-)$ of the d which it represents.
4. Determine $T =$ the smaller of the sums of the like-signed ranks.
5. By counting, determine $N =$ the total number of d's having a sign.
6. The procedure for determining the significance of the observed value of T depends on the side of N:

   a. If N is 25 or less, Table G shows critical values of T for various sizes of N. If the observed value of T is equal to or less than that given in the table for a particular significance level and a particular N, H_0 may be rejected at that level of significance.
   b. If N is larger than 25, compute the value of z as defined by formula (5.5). Determine its associated probability under H_0 by referring to Table A. For a two-tailed test, double the p shown.

If the p thus obtained is equal to or less than α, reject H_0.

Power-Efficiency

When the assumptions of the parametric t test (see page 19) are in fact met, the asymptotic efficiency near H_0 of the Wilcoxon matched-pairs signed-ranks test compared with the t test is $3/\pi = 95.5$ per cent (Mood, 1954). This means that $3/\pi$ is the limiting ratio of sample sizes necessary for the Wilcoxon test and the t test to attain the same power. For small samples, the efficiency is near 95 per cent.

References

The reader may find other discussions of the Wilcoxon matched-pairs signed-ranks test in Mood (1954), Moses (1952a), and Wilcoxon (1945; 1947; 1949).

THE WALSH TEST

Function

If the experimenter can assume that the difference scores he observes in two related samples are drawn from symmetrical populations, he may use the very powerful test developed by Walsh. Notice that the assumption is *not* that the d_i's are from normal populations (which is the assumption of the parametric t test), and notice that the d_i's do not even have to be from the same population. What the test does assume is that the populations are symmetrical, so that the mean is an accurate representation of central tendency, and is equal to the median. The Walsh test requires measurement in at least an interval scale.

Method

To use the Walsh test, one first obtains difference scores (d_i's) for each of the N pairs. These d_i's are then arranged in order of size, with the sign of each d taken into consideration in this arrangement. Let $d_1 =$ the lowest difference score (this may well be a negative d), $d_2 =$ the next lowest difference, etc. Thus $d_1 \leq d_2 \leq d_3 \leq d_4 \leq \cdots \leq d_N$.

The null hypothesis to be tested is that these d_i's were drawn from a population whose median $= 0$ (or from a group of populations whose common median $= 0$). In a symmetrical distribution, the mean and the median coincide. The Walsh test assumes that the d_i's are from populations with symmetrical distributions. Therefore H_0 is that the average of the difference scores (μ_0) is zero. For a two-tailed test, H_1 is that $\mu_1 \neq 0$. For a one-tailed test, H_1 may be either that $\mu_1 > 0$, or that $\mu_1 < 0$.

Table H of the Appendix is used to determine the significance of various results under Walsh's test. To use this table, one must know the observed value of N (the number of pairs), the nature of H_1, and the numerical values of every d_i.

Table H gives significant values for both one-tailed and two-tailed tests. The two right-hand columns give the values which permit rejecting H_0 at the stated significance level. If H_1 is that $\mu_1 \neq 0$, then the null hypothesis may be rejected if either of the tabled values are observed. If H_1 is directional, then the null hypothesis may be rejected if the values tabled under that H_1 are observed.

The left-hand column shows various values of N, from 4 to 15. Next to that column are two columns which show the significance levels at which the tabled values may be rejected.

Since Table H is somewhat more complicated than most, we will give several examples of its use.

Suppose $N = 5$. For a two-tailed test, we may reject H_0 at the $\alpha = .125$ level if $\frac{1}{2}(d_4 + d_5)$ is less than zero *or* if $\frac{1}{2}(d_1 + d_2)$ is larger than zero. And we may reject H_0 at $\alpha = .062$ if d_5 is less than zero *or* if d_1 is larger than zero.

Now suppose that $N = 5$ and that we had predicted in advance that our difference scores would be larger than zero. Then if $\frac{1}{2}(d_1 + d_2)$ is larger than zero, we can reject H_0 at $\alpha = .062$. And if d_1 is larger than zero, we can reject H_0 at $\alpha = .031$.

On the other hand, suppose we had predicted in advance that our difference scores would be negative. That is, H_1 is that $\mu_1 < 0$. Then, if $N = 5$, if $\frac{1}{2}(d_4 + d_5)$ were less than zero we could reject H_0 at $\alpha = .062$. And if d_5 were less than zero, we could reject H_0 at $\alpha = .031$.

Now for larger values of N, Table H is somewhat more complicated. The two right-hand columns give alternative values, the alternatives being separated by a comma. Accompanying these are "max" and "min." "Max" means that we should select that alternative which is larger; "min" means that we should select that alternative which is smaller.

For example, at $N = 6$, suppose H_1 is that $\mu_1 < 0$. We may reject H_0 in favor of that H_1 at $\alpha = .047$ if d_5 or $\frac{1}{2}(d_4 + d_6)$, *whichever is larger*, is less than zero.

In the example of the application of this test given below, the use of Table H is illustrated again.

Example

In a study designed to induce repression, Lowenfeld[1] had his fifteen subjects learn 10 nonsense syllables. He then attempted to associate negative affect to 5 of these (selected at random from the 10) by giving the subjects an electric shock whenever any one of the 5 syllables was exposed tachistoscopically. After a lapse of 48 hours, the subjects were brought back to the experimental room and asked to recall the list of nonsense syllables. The prediction was that they would recall more of the nonshock syllables than the shock syllables.

i. *Null Hypothesis.* H_0: the median difference between the number of nonshock syllables remembered and the number of shock syllables remembered is zero. That is, subjects will recall the two groups of syllables equally well. H_1: the number of nonshock syllables remembered is larger than the number of shock syllables remembered. That is, the median difference will be larger than zero.

ii. *Statistical Test.* The Walsh test was chosen because the study uses two related samples (each subject serving as his own control), and because the assumption that the numerical difference scores came from symmetrical populations seemed tenable.

iii. *Significance Level.* Let $\alpha = .05$. $N = 15 =$ the number of subjects who served in the study, each being exposed to both shock and nonshock syllables.

iv. *Sampling Distribution.* Table H gives the associated proba-

---

[1] Lowenfeld, J. 1955. An experiment relating the concepts of repression, subception, and perceptual defense. Unpublished doctor's dissertation, The Pennsylvania State University.

bility of occurrence under H_0 for various values of the statistical test when $N \leq 15$.

v. *Rejection Region.* Since the direction of the differences was predicted in advance, a one-tailed region of rejection will be used. Since H_1 is that $\mu_1 > 0$, H_0 will be rejected if any of the values given in the right-hand column of the table for $N = 15$ should occur, since the levels of significance for all of the values tabled for $N = 15$ are less than $\alpha = .05$.

vi. *Decision.* The number of shock and nonshock syllables recalled by each subject after 48 hours is given in Table 5.8, which

TABLE 5.8. NUMBER OF SHOCK AND NONSHOCK SYLLABLES RECALLED
AFTER 48 HOURS

Subject	Number of nonshock syllables recalled	Number of shock syllables recalled	d
a	5	2	3
b	4	2	2
c	3	0	3
d	5	3	2
e	2	3	-1
f	4	2	2
g	2	3	-1
h	2	1	1
i	4	1	3
j	4	3	1
k	3	4	-1
l	1	2	-1
m	5	2	3
n	3	4	-1
o	1	0	1

also gives the d for each. Thus subject a recalled 5 of the nonshock syllables but only 2 of the shock syllables; his $d = 5 - 2 = 3$.

Notice that the smallest d is -1. Thus $d_1 =$ the lowest d, taking sign into consideration $= -1$. Five of the d's are -1's; therefore $d_1 = -1$, $d_2 = -1$, $d_3 = -1$, $d_4 = -1$, and $d_5 = -1$.

The next smallest d's are 1's. Three subjects (h, j, and o) have d's of 1. Therefore $d_6 = 1$, $d_7 = 1$, and $d_8 = 1$.

Three of the d's are 2's. Thus $d_9 = 2$, $d_{10} = 2$, and $d_{11} = 2$.

The largest d's are 3's. There are four of them. Thus $d_{12} = 3$, $d_{13} = 3$, $d_{14} = 3$, and $d_{15} = 3$.

Now Table H shows that for $N = 15$, the one-tailed test for the H_1 that $\mu_1 > 0$ at $\alpha = .047$ is

$$\text{Minimum} \left[\tfrac{1}{2}(d_1 + d_{12}), \tfrac{1}{2}(d_2 + d_{11}) \right] > 0$$

The "minimum" means that we should choose the smaller of the two values given, in terms of our observed values of d. That is, if $\frac{1}{2}(d_1 + d_{12})$ *or* $\frac{1}{2}(d_2 + d_{11})$, whichever is smaller, is larger than zero, then we may reject H_0 at $\alpha = .047$.

As we have shown, $d_1 = -1$, $d_{12} = 3$, $d_2 = -1$, and $d_{11} = 2$. Substituting these values, we have

$$\text{Minimum } [\tfrac{1}{2}(-1 + 3), \tfrac{1}{2}(-1 + 2)]$$
$$= \text{minimum } [\tfrac{1}{2}(2), \tfrac{1}{2}(1)]$$
$$= \tfrac{1}{2}(1)$$

We see that for our data the smaller of these two values is $\frac{1}{2}(1) = \frac{1}{2}$. Since this value *is* larger than zero, we can reject H_0 at $\alpha = .047$. Since the probability under H_0 associated with the values which occurred is less than $\alpha = .05$, we decide to reject H_0 in favor of H_1.[*] We conclude that the number of nonshock syllables remembered was significantly larger than the number of shock syllables remembered, a conclusion which supports the theory that negative affect induces repression.

Summary of procedure. These are the steps in the use of the Walsh test:

1. Determine the signed difference score (d_i) for each matched pair.
2. Determine N, the number of matched pairs.
3. Arrange the d_i's in order of increasing size, from d_1 to d_N. Take the sign of the d into account in this ordering. Thus d_1 is the largest negative d, and d_N is the largest positive d.
4. Consult Table H to determine whether H_0 may be rejected in favor of H_1 with the observed values of $d_1, d_2, d_3, \ldots, d_N$. The technique of using Table H is explained above at some length.

Power-Efficiency

When compared with the most powerful test, the parametric t test, the Walsh test has power-efficiency (in the sense defined in Chap. 3) of 95 per cent for most values of N and α. Its power-efficiency is as high as 99 per cent (for $N = 9$ and $\alpha = .01$, one-tailed test) and is nowhere lower than 87.5 per cent (for $N = 10$ and $\alpha = .06$, one-tailed test). For information on its power-efficiency, see Walsh (1949b).

References

For other discussions of the Walsh test, the reader is referred to Dixon and Massey (1951, chap. 17) and to Walsh (1949a; 1949b).

---

[*] Using the nonparametric Wilcoxon matched-pairs signed-ranks test, **Lowenfeld** came to the same decision.

THE RANDOMIZATION TEST FOR MATCHED PAIRS
Function

Randomization tests are nonparametric tests which not only have practical value in the analysis of research data but also have heuristic value in that they help expose the underlying nature of nonparametric tests in general. With a randomization test, we can obtain the exact probability under H_0 associated with the occurrence of our observed data, and we can do this without making any assumptions about normality or homogeneity of variance. Randomization tests, under certain conditions, are the most powerful of the nonparametric techniques, and may be used whenever measurement is so precise that the values of the scores have numerical meaning.

Rationale and Method

Consider the small sample example to which we earlier applied the Wilcoxon matched-pairs signed-ranks test (discussed on pages 77 to 78). In that study, we had 8 matched pairs, and one member of each pair was randomly assigned to each condition—one twin attended nursery school while the other stayed at home. The research hypothesis predicted differences between these two groups in "social perceptiveness" because of the different treatment conditions. The null hypothesis was that the two conditions produced no difference in social perceptiveness. It will be remembered that the two members of any matched pair were assigned to the conditions by some random method, say by tossing a coin.

For this discussion, let us assume that in the fictitious research under discussion measurement was achieved in the sense of an interval scale.

Now if the null hypothesis that there is no treatment effect were really true, we would have obtained the same social perceptiveness scores if both groups had attended the nursery school or if both groups had stayed at home. That is, under H_0 these children would have scored as they did regardless of the conditions. We may not know why the children differ among themselves in social perceptiveness, but under H_0 we do know how the *signs* of the difference scores arose: they resulted from the random assignment of the children to the two conditions. For example, for the two twins in pair a we observed a difference of 19 points between their two scores in social perceptiveness. Under H_0, we presume that this d was $+19$ rather than -19 simply because we happened to assign to the nursery school group that twin who would have been higher in social perceptiveness anyway. The d was $+19$ rather than -19 simply because when we were assigning the twins to treatments our coin fell on head rather than on tail. By this reasoning, under H_0 every difference score we observed could equally likely have had the opposite sign.

The difference scores that we observed in our sample in that study happened to be

$$+19 \quad +27 \quad -1 \quad +6 \quad +7 \quad +13 \quad -4 \quad +3$$

Under H_0, if our coin tosses had been different, they might just as probably have been

$$-19 \quad -27 \quad +1 \quad -6 \quad -7 \quad -13 \quad +4 \quad -3$$

or if the coins had fallen still another way they would have been

$$+19 \quad -27 \quad +1 \quad -6 \quad -7 \quad -13 \quad -4 \quad +3$$

As a matter of fact, if the null hypothesis is true, then there are $2^N = 2^8$ equally likely outcomes, and the one which we observe depends entirely on how the coin landed for each of the 8 tosses when we assigned the twins to the two groups.   This means that associated with the sample of scores we observed there are many other possible ones, the total possible combinations being $2^8 = 256$.   Under H_0, any one of these 256 possible outcomes was just as likely to occur as the one which did occur.

For each of the possible outcomes there is a sum of the differences: Σd_i.   Now many of the 256 Σd_i are near zero, about what we should expect if H_0 were true.   A few Σd_i are far from zero.   These are for those combinations in which nearly all of the signs are plus or are minus.   It is such combinations which we should expect if the population mean under one of the treatments exceeds that under the other, that is, if H_0 is *false*.

If we wish to test H_0 against some H_1, we set up a region of rejection consisting of the combinations whose Σd_i is largest.   Suppose $\alpha = .05$. Then the region of rejection consists of that 5 per cent of the possible combinations which contains the most extreme values of Σd_i.

In the example under discussion, 256 possible outcomes are equally likely under H_0.   The region of rejection consists of the 12 most extreme possible outcomes, for $(.05)(256) = 12.8$.   Under the null hypothesis, the probability that we will observe one of these 12 extreme outcomes is $\frac{12}{256} = .047$.   If we actually observe one of those extreme outcomes which is included in the region of rejection, we reject H_0 in favor of H_1.

When a one-tailed test is called for, the region of rejection consists of the same number of samples.   However, it consists of that number of the most extreme possible outcomes in one direction, either positive or negative, depending on the direction of the prediction in H_1.

When a two-tailed test is called for, as is the case in the example under discussion, the region of rejection consists of the most extreme possible outcomes at both the positive and the negative ends of the distribution of Σd_i's.   That is, in the example, the 12 outcomes in the region of rejection would include the 6 yielding the largest positive Σd_i and the 6 yielding the largest negative Σd_i.

Example

i. *Null Hypothesis.* H_0: the two treatments are equivalent. That is, there is no difference in social perceptiveness under the two conditions (attendance at nursery school or staying at home). In social perceptiveness, all 16 observations (8 pairs) are from a common population. H_1: the two treatments are not equivalent.

ii. *Statistical Test.* The randomization test for matched pairs is chosen because of its appropriateness to this design (two related samples, N not cumbersomely large) and because for these (artificial) data we are willing to consider that its requirement of measurement in at least an interval scale is met.

iii. *Significance Level.* Let $\alpha = .05$. $N =$ the number of pairs $= 8$.

iv. *Sampling Distribution.* The sampling distribution consists of the permutation of the signs of the differences to include all possible (2^N) occurrences of Σd_i. In this case, $2^N = 2^8 = 256$.

v. *Rejection Region.* Since H_1 does not predict the direction of the differences, a two-tailed test is used. The region of rejection consists of those 12 outcomes which have the most extreme Σd_i's, 6 being the most extreme positive Σd_i's and 6 being the most extreme negative Σd_i's.

vi. *Decision.* The data of this study are shown in Table 5.6. The d's observed were:

$$+19 \quad +27 \quad -1 \quad +6 \quad +7 \quad +13 \quad -4 \quad +3$$

For these d's, $\Sigma d_i = +70$.

Table 5.9 shows the 6 possible outcomes with the most extreme

TABLE 5.9. THE SIX MOST EXTREME POSSIBLE POSITIVE OUTCOMES
FOR THE d'S SHOWN IN TABLE 5.6
(These constitute one tail of the rejection region for the randomization
test when $\alpha = .05$)

	Outcome								Σd_i
(1)	+19	+27	+1	+6	+7	+13	+4	+3	80
(2)	+19	+27	-1	+6	+7	+13	+4	+3	78
(3)	+19	+27	+1	+6	+7	+13	+4	-3	74
(4)	+19	+27	+1	+6	+7	+13	-4	+3	72
(5)	+19	+27	-1	+6	+7	+13	+4	-3	72
(6)*	+19	+27	-1	+6	+7	+13	-4	+3	70

Σd_i's at the positive end of the sampling distribution. These 6 outcomes constitute one tail of the two-tailed region of rejection for

$N = 8$. Outcome 6 (with an asterisk) is the outcome we actually observed. The probability of its occurrence or a set more extreme under H_0 is $p = .047$. Since this p is less than $\alpha = .05$, our decision in this fictitious study is to reject the null hypothesis of no condition differences.

Large samples. If the number of pairs exceeds, say, $N = 12$, the randomization test becomes unwieldy. For example, if $N = 13$, the number of possible outcomes is $2^{13} = 8,192$. Thus the region of rejection for $\alpha = .05$ would consist of $(.05)(8,192) = 409.6$ possible extreme outcomes. The computation necessary to specify the region of rejection would therefore be quite tedious.

Because of the computational cumbersomeness of the randomization test when N is at all large, it is suggested that the Wilcoxon matched-pairs signed-ranks test be used in such cases. In the Wilcoxon test, ranks are substituted for numbers. It provides a very efficient alternative to the randomization test because it is in fact a randomization test on the ranks.[1] Even if we did not have the use of Table G, it would not be too tedious to compute the test by permuting the signs ($+$ and $-$) on the set of ranks in all possible ways and then tabulating the upper and lower significance points for a given sample size.

If N is larger than 25, and if the differences show little variability, another alternative is available. If the d_i be all about the same size, so that $\dfrac{d_{max}^2}{\Sigma d_i^2} \leq \dfrac{5}{2N}$, where d_{max}^2 is the square of the largest observed difference, then the central-limit theorem (see Chap. 2) may be expected to hold (Moses, 1952a). Under these conditions, we can expect Σd_i to be approximately normally distributed with

$$\text{Mean} = 0$$

and

$$\text{Standard deviation} = \sqrt{\Sigma d_i^2}$$

and therefore

$$z = \frac{\Sigma d_i - \mu}{\sigma} = \frac{\Sigma d_i}{\sqrt{\Sigma d_i^2}} \tag{5.6}$$

is approximately normally distributed with zero mean and unit variance. Table A of the Appendix gives the probability associated with the occur-

---

[1] In a randomization test on ranks, all 2^N permutations of the signs of the ranks are considered, and the most extreme possible constitute the region of rejection. For the data shown in Table 5.6, there are $2^8 = 256$ possible and equally likely combinations of signed ranks under H_0. The curious reader can satisfy himself that the sample of signed ranks observed is among the 12 most extreme possible outcomes and thus leads us to reject H_0 at $\alpha = .05$, which was our decision which we based on Table G. By this randomization method, Table G, the table of significant values of T, can be reconstructed.

rence under H_0 of values as extreme as any z obtained through the application of formula (5.6).

However, the requirement that the d_i's show little variability, i.e., that $\dfrac{d_{max}^2}{\Sigma d_i^2} \leq \dfrac{5}{2N}$, is not too commonly met. For this reason, and also because the efficiency of the Wilcoxon test (approximately 95 per cent for large samples) is very likely to be superior to that of this large sample approximation to the randomization test when nonnormal populations are involved, it would seem that the Wilcoxon test is the better alternative when N's are cumbersomely large.

Summary of procedure. When N is small and when measurement is in at least an interval scale, the randomization test for matched pairs may be used. These are the steps:

1. Observe the values of the various d_i's and their signs.

2. Determine the number of possible outcomes under H_0 for these values: 2^N.

3. Determine the number of possible outcomes in the region of rejection: $(\alpha)(2^N)$.

4. Identify those possible outcomes which are in the region of rejection by choosing from the possible outcomes those with the largest Σd_i's. For a one-tailed test, the outcomes in the region of rejection are all in one direction (either positive or negative). For a two-tailed test, half of the outcomes in the region of rejection are those with the largest positive Σd_i's and half are those with the largest negative Σd_i's.

5. Determine whether the observed outcome is one of those in the region of rejection. If it is, reject H_0 in favor of H_1.

When N is large, the Wilcoxon matched-pairs signed-ranks test is recommended for use rather than the randomization test. When N is 25 or larger and when the data meet certain specified conditions, an approximation [formula (5.6)] may also be used.

Power-Efficiency

The randomization test for matched pairs, because it uses all of the information in the sample, has power-efficiency of 100 per cent.

References

Discussions of the randomization method are contained in Fisher (1935), Moses (1952a), Pitman (1937a; 1937b; 1937c), Scheffé (1943), and Welch (1937).

DISCUSSION

In this chapter we have presented five nonparametric statistical tests for the case of two related samples (the design in which matched pairs are used). The comparison and contrast of these tests which are presented below may aid the reader in choosing from among these tests that one which will be most appropriate to the data of a particular experiment.

All the tests but the McNemar test for the significance of changes assume that the variable under consideration has a continuous distribution underlying the scores. Notice that there is no requirement that the measurement itself be continuous; the requirement concerns the variable of which the measurement gives some gross or approximate representation.

The McNemar test for the significance of changes may be used when one or both of the conditions under study has been measured only in the sense of a nominal scale. For the case of two related samples, the McNemar test is unique in its suitability for such data. That is, this test should be used when the data are in frequencies which can only be classified by separate categories which have no relation to each other of the "greater than" type. No assumption of a continuous variable need be made, because this test is equivalent to a test by the binomial distribution with $P = Q = \frac{1}{2}$, where $N =$ the number of changes.

If ordinal measurement within pairs is possible (i.e., if the score of one member of a pair can be ranked as "greater than" the score of the other member of the same pair), then the sign test is applicable. That is, this test is useful for data on a variable which has underlying continuity but which can be measured in only a very gross way. When the sign test is applied to data which meet the conditions of the parametric alternative (the t test), it has power-efficiency of about 95 per cent for $N = 6$, but its power-efficiency declines as N increases to about 63 per cent for very large samples.

When the measurement is in an ordinal scale both *within* and *between* pairs, the Wilcoxon test should be used. That is, it is applicable when the researcher can meaningfully rank the differences observed for the various matched pairs. It is not uncommon for behavioral scientists to be able to rank difference scores in the order of their absolute size without being able to give truly numerical scores to the observations in each pair. When the Wilcoxon test is used for data which in fact meet the conditions of the t test, its power-efficiency is about 95 per cent for large samples and not much less than that for smaller samples.

If the experimenter can assume that the populations from which he has sampled are both symmetrical and continuous, then the Walsh test is applicable when N is 15 or less. This test requires measurement in at least an interval scale. It has power-efficiency (in the sense previously defined) of about 95 per cent for most values of N and α.

The randomization test should be used whenever N is sufficiently small to make it computationally feasible and when the measurement of the variable is at least in an interval scale. The randomization test uses all the information in the sample and thus is 100 per cent efficient on data which may properly be analyzed by the t test.

Of course none of these nonparametric tests makes the assumption of normality which is made by the comparable parametric test, the t test.

In summary, we conclude that the McNemar test for the significance of changes should be used for both large and small samples when the measurement of at least one of the variables is merely nominal. For the crudest of ordinal measurement, the sign test should be used. For more refined measurement, the Wilcoxon matched-pairs signed-ranks test may be used in all cases. For N's of 15 or fewer, the Walsh test may be used. If interval measurement is achieved, the randomization test should be used when the N is not so large as to make its computation cumbersome.

CHAPTER 6

THE CASE OF TWO INDEPENDENT SAMPLES

In studying differences between two groups, we may use either related or independent groups. Chapter 5 offered statistical tests for use in a design having two related groups. The present chapter presents statistical tests for use in a design having two independent groups. Like those presented in Chap. 5, the tests presented here determine whether differences in the samples constitute convincing evidence of a difference in the processes applied to them.

Although the merits of using two related samples in a research design are great, to do so is frequently impractical. Frequently the nature of the dependent variable precludes using the subjects as their own controls, as is the case when the dependent variable is length of time in solving a particular unfamiliar problem. A problem can be unfamiliar only once. It may also be impossible to design a study which uses matched pairs, perhaps because of the researcher's ignorance of useful matching variables, or because of his inability to obtain adequate measures (to use in selecting matched pairs) of some variable known to be relevant, or finally because good "matches" are simply unavailable.

When the use of two related samples is impractical or inappropriate, one may use two independent samples. In this design the two samples may be obtained by either of two methods: (a) they may each be drawn at random from two populations, or (b) they may arise from the assignment at random of two treatments to the members of some sample whose origins are arbitrary. In either case it is not necessary that the two samples be of the same size.

An example of random sampling from two populations would be the drawing of every tenth Democrat and every tenth Republican from an alphabetical list of registered voters. This would result in a random sample of registered Democrats and Republicans from the voting area covered by the list, and the number of Democrats would equal the number of Republicans only if the registration of the two parties happened to be substantially equal in that area. Another example would be the drawing of every eighth upperclassman and every twelfth lowerclassman from a list of students in a college.

95

An example of the random assignment of method might occur in a study of the effectiveness of two instructors in teaching the same course. A registration card might be collected from every student enrolled in the course, and at random one half of these cards would be assigned to one instructor and one half to the other.

The usual parametric technique for analyzing data from two independent samples is to apply a t test to the means of the two groups. The t test assumes that the scores (which are summed in the computing of the means) are independent observations from normally distributed populations with equal variances. This test, because it uses means and other statistics arrived at by arithmetical computation, requires that the observations be measured on at least an interval scale.

For a given research, the t test may be inapplicable for a variety of reasons. The researcher may find that (a) the assumptions of the t test are unrealistic for his data, (b) he prefers to avoid making the assumptions and thus to give his conclusions greater generality, or (c) his "scores" may not be truly *numerical* and therefore fail to meet the measurement requirement of the t test. In instances like these, the researcher may choose to analyze his data with one of the nonparametric statistical tests for two independent samples which are presented in this chapter. The comparison and contrast of these tests in the discussion at the conclusion of the chapter may aid him in choosing from among the tests presented that one which is best suited for the data of his study.

THE FISHER EXACT PROBABILITY TEST

Function

The Fisher exact probability test is an extremely useful nonparametric technique for analyzing discrete data (either nominal or ordinal) when the two independent samples are small in size. It is used when the scores from two independent random samples all fall into one or the other of two mutually exclusive classes. In other words, every subject in both groups obtains one of two possible scores. The scores are represented by frequencies in a 2 × 2 contingency table, like Table 6.1. Groups I and II might be any two independent groups, such as experimentals and controls,

TABLE 6.1. 2 × 2 CONTINGENCY TABLE

	−	+	Total
Group I	A	B	$A + B$
Group II	C	D	$C + D$
Total	$A + C$	$B + D$	N

males and females, employed and unemployed, Democrats and Republicans, fathers and mothers, etc. The column headings, here arbitrarily indicated as plus and minus, may be any two classifications: above and below the median, passed and failed, science majors and arts majors, agree and disagree, etc. The test determines whether the two groups differ in the proportion with which they fall into the two classifications. For the data in Table 6.1 (where A, B, C, and D stand for frequencies) it would determine whether Group I and Group II differ significantly in the proportion of pluses and minuses attributed to them.

Method

The exact probability of observing a particular set of frequencies in a 2×2 table, when the marginal totals are regarded as fixed, is given by the hypergeometric distribution

$$p = \frac{\binom{A + C}{A}\binom{B + D}{B}}{\binom{N}{A + B}}$$

$$= \frac{\left(\frac{(A + C)!}{A!\,C!}\right)\left(\frac{(B + D)!}{B!\,D!}\right)}{\frac{N!}{(A + B)!\,(C + D)!}}$$

and thus
$$p = \frac{(A + B)!\,(C + D)!\,(A + C)!\,(B + D)!}{N!\,A!\,B!\,C!\,D!} \tag{6.1}$$

That is, the exact probability of the observed occurrence is found by taking the ratio of the product of the factorials of the four marginal totals to the product of the cell frequencies multiplied by N factorial. (Table S of the Appendix may be helpful in these computations.)

To illustrate the use of formula (6.1): suppose we observe the data shown in Table 6.2. In that table, $A = 10$, $B = 0$, $C = 4$, and $D = 5$. The marginal totals are $A + B = 10$, $C + D = 9$, $A + C = 14$, and $B + D = 5$. N, the total number of independent observations, is 19. The exact probability that these 19 cases should fall in the four cells as

TABLE 6.2

	−	+	Total
Group I	10	0	10
Group II	4	5	9
Total	14	5	19

they did may be determined by substituting the observed values in formula (6.1):

$$p = \frac{10!\,9!\,14!\,5!}{19!\,10!\,0!\,4!\,5!}$$

$$= .0108$$

We determine that the probability of such a distribution of frequencies under H_0 is $p = .0108$.

Now the above example was a comparatively simple one to compute because one of the cells (cell B) had a frequency of 0. But if none of the cell frequencies is zero, we must remember that more extreme deviations from the distribution under H_0 could occur with the same marginal totals, and we must take into consideration these possible "more extreme" deviations, for a statistical test of the null hypothesis asks: What is the probability under H_0 of such an occurrence *or of one even more extreme?*

For example, suppose the data from a particular study were those given in Table 6.3. With the marginal totals unchanged, a more extreme

TABLE 6.3

	−	+	Total
Group I	1	6	7
Group II	4	1	5
Total	5	7	12

occurrence would be that shown in Table 6.4. Thus, if we wish to apply

TABLE 6.4

	−	+	Total
Group I	0	7	7
Group II	5	0	5
Total	5	7	12

a statistical test of the null hypothesis to the data given in Table 6.3, we must sum the probability of that occurrence with the probability of the more extreme possible one (shown in Table 6.4). We compute each p by using formula (6.1). Thus we have

$$p = \frac{7!\,5!\,5!\,7!}{12!\,1!\,6!\,4!\,1!}$$

$$= .04399$$

and

$$p = \frac{7!\,5!\,5!\,7!}{12!\,0!\,7!\,5!\,0!}$$

$$= .00126$$

Thus the probability of the occurrence in Table 6.3 or of an even more extreme occurrence (shown in Table 6.4) is

$$p = .04399 + .00126$$
$$= .04525$$

That is, $p = .04525$ is the value of p which we use in deciding whether the data in Table 6.3 permit us to reject H_0.

The reader can readily see that if the smallest cell value in the contingency table is even moderately large, the Fisher test becomes computationally very tedious. For example, if the smallest cell value is 2, then three exact probabilities must be determined by formula (6.1) and then summed; if the smallest cell value is 3, then four exact probabilities must be found and summed, etc.

If the researcher is content to use significance levels rather than exact values of p, Table I of the Appendix may be used. It eliminates the necessity for the tedious computations illustrated above. Using it, the researcher may determine directly the significance of an observed set of values in a 2 $\times$ 2 contingency table. Table I is applicable to data where N is 30 or smaller, and where neither of the totals in the right-hand margin is larger than 15. That is, neither $A + B$ nor $C + D$ may be larger than 15. (The researcher may find that the bottom marginal totals in his data meet this requirement but the right-hand totals do not. Obviously, in that case he may meet the requirement by simply recasting the data, i.e., by shifting the labels at the top of the contingency table to the left margin, and vice versa.)

Because of its very size, Table I is somewhat more difficult to use than are most tables of significance values. Therefore we include detailed directions for its use. These are the steps in the use of Table I:

1. Determine the values of $A + B$ and $C + D$ in the data.

2. Find the observed value of $A + B$ in Table I under the heading "Totals in Right Margin."

3. In that section of the table, locate the observed value of $C + D$ under the same heading.

4. For the observed value of $C + D$, several possible values of B^* are listed in the table. Find the observed value of B among these possibilities.

5. Now observe your value of D. If the observed value of D is equal to or less than the value given in the table under your level of significance, then the observed data are significant at that level.

It should be noted that the significance levels given in Table I are approximate. And they err on the conservative side. Thus the exact probability of some data may be $p = .007$, but Table I will show it signifi-

* If the observed value of B is not included among them, use the observed value of A instead. If A is used in place of B, then C is used in place of D in step 5.

cant at $\alpha = .01$. If the reader requires exact probabilities rather than significance levels, he may find these in Finney (1948, pp. 145–156) or he may compute them by using formula (6.1) in the manner described earlier.

Notice also that the levels of significance given in Table I are for one-tailed regions of rejection. If a two-tailed rejection region is called for, double the significance level given in Table I.

The reader's understanding of the use of Table I may be aided by an example. We recur to the data given in Table 6.3, for which we have already determined the exact probability by using formula (6.1). For Table 6.3, $A + B = 7$ and $C + D = 5$. The reader may find the appropriate section in Table I for such right marginal totals. In that section he will find that three alternative values of B (7, 6, and 5) are tabled. Now in Table 6.3, $B = 6$. Therefore the reader should use the middle of the three lines of values, that in which $B = 6$. Now observe the value of D in our data: $D = 1$ in Table 6.3. Table I shows that $D = 1$ is significant at the .05 level (one-tailed). This agrees with the exact probability we computed: $p = .045$.

For a two-tailed test we would double the observed significance level, and conclude that the data in Table 6.3 permit us to reject H_0 at the $\alpha = 2(.05) = .10$ level.

Example

In a study of the personal and social backgrounds of the leaders of the Nazi movement, Lerner and his collaborators[1] compared the Nazi elite with the established and respected elite of the older German society. One such comparison concerned the career histories of the 15 men who constituted the German Cabinet at the end of 1934. These men were categorized in two groups: Nazis and non-Nazis. To test the hypothesis that Nazi leaders had taken political party work as their careers while non-Nazis had come from other, more stable and conventional, occupations, each man was categorized according to his first job in his career. The first job of each was classified as either "stable occupation" or as "party administration and communication." The hypothesis was that the two groups would differ in the proportion with which they were assigned to these two categories.

 i. *Null Hypothesis.* H_0: Nazis and non-Nazis show equal proportions in the kind of "first jobs" they had. H_1: a greater proportion of Nazis' "first jobs" were in party administration and communication than were the "first jobs" of non-Nazi politicians.

 ii. *Statistical Test.* This study calls for a test to determine the significance of the difference between two independent samples.

---

[1] Lerner, D., Pool, I. de S., and Schueller, G. K. 1951. *The Nazi elite.* Stanford, Calif.: Stanford Univer. Press. The data cited in this example are given on p. 101.

Since the measures are both dichotomous and since N is small, the Fisher test is selected.

iii. *Significance Level.* Let $\alpha = .05$. $N = 15$.

iv. *Sampling Distribution.* The probability of the occurrence under H_0 of an observed set of values in a 2×2 table may be found by the use of formula (6.1). However, for $N \leq 30$ (which is the case with these data), Table I may be used. It gives critical values of D for various levels of significance.

v. *Rejection Region.* Since H_1 predicts the direction of the difference, the region of rejection is one-tailed. H_0 will be rejected if the observed call values differ in the predicted direction and if they are of such magnitude that the probability associated with their occurrence under H_0 is equal to or less than $\alpha = .05$.

vi. *Decision.* The information concerning the "first jobs" of each member of the German Cabinet late in 1934 is given in Table 6.5. For this table, $A + B = 9$ and $C + D = 6$. Reference to

TABLE 6.5. FIELD OF FIRST JOB OF 1934 MEMBERS OF
GERMAN CABINET

	Stable occupations (law and civil service)	Party administration and communication	Total
Nazis	1	8	9
Non-Nazis	6	0	6
Total	7	8	15

Table I reveals that with these marginal totals, and with $B = 8$, the observed $D = 0$ has a one-tailed probability of occurrence under H_0 of $p \leq .005$. Since this p is smaller than our level of significance, $\alpha = .05$, our decision is to reject H_0 in favor of H_1. We conclude that Nazi and non-Nazi political leaders did differ in the fields of their first jobs.[1]

Tocher's modification. In the literature of statistics, there has been considerable discussion of the applicability of the Fisher test to various sorts of data, inasmuch as there seems to be something arbitrary or improper about considering the marginal totals fixed, for the marginal totals might easily vary if we actually drew repeated samples of the same size by the same method from the same population. Fisher (1934) recommends the test for all types of dichotomous data, but this recommendation has been questioned by others.

---

[1] Lerner *et al.* come to the same conclusion, although they do not report any statistical test of these data.

However, Tocher (1950) has proved that a slight modification of the Fisher test provides the most powerful one-tailed test for data in a 2 × 2 table. We will illustrate this modification by giving Tocher's example. Table 6.6 shows some observed frequencies (in a) and shows the two more

TABLE 6.6. TOCHER'S EXAMPLE

Observed data	More extreme outcomes with same marginal totals	
a	b	c

2	5	7
3	2	5
5	7	12

1	6	7
4	1	5
5	7	12

0	7	7
5	0	5
5	7	12

extreme distributions of frequencies which could occur with the same marginal totals (b and c). Given the observed data (a), we wish to test H_0 at $\alpha = .05$. Applying formula (6.1) to the data in each of the three tables, we have

$$p_a = \frac{7!\,5!\,5!\,7!}{12!\,2!\,5!\,3!\,2!} = .26515$$

$$p_b = \frac{7!\,5!\,5!\,7!}{12!\,1!\,6!\,4!\,1!} = .04399$$

$$p_c = \frac{7!\,5!\,5!\,7!}{12!\,0!\,7!\,5!\,0!} = .00126$$

The probability associated with the occurrence of values as extreme as the observed scores (a) under H_0 is given by adding these three p's:

$$.26515 + .04399 + .00126 = .31040$$

Thus $p = .31040$ is the probability we would find by the Fisher test.

Tocher's modification first determines the probability of all the cases more extreme than the observed one, and not including the observed one. Thus in this case one would sum only p_b and p_c:

$$.04399 + .00126 = .04525$$

Now if this probability of the more extreme outcomes is larger than α, we cannot reject H_0. But if this probability is less than α while the probability yielded by the Fisher test is greater than α (as is the case with these data), then Tocher recommends computing this ratio:

$$\frac{\alpha - p_{\text{more extreme cases}}}{p_{\text{observed case taken alone}}} \tag{6.2}$$

For the data shown in Table 6.6, this would be

$$\frac{\alpha - (p_b + p_c)}{p_a}$$

which is

$$\frac{.05 - .0425}{.26515} = .01791$$

Now we go to a table of random numbers and at random draw a number between 0 and 1. If this random number is *smaller* than our ratio above (i.e., if it is smaller than .01791), we reject H_0. If it is larger, we cannot reject H_0. Of course in this case it is highly unlikely that the randomly drawn number will be sufficiently small to permit us to reject H_0. But this added small probability of rejecting H_0 makes the Fisher test slightly less conservative.

Perhaps the reader will gain an intuitive understanding of the logic and power of Tocher's modification by considering what a one-tailed test at $\alpha = .05$ really is for the data given in Table 6.6. Suppose we reject H_0 only when cases b or c occur. Then we are actually working at $\alpha = .04525$. In order to move to exactly the $\alpha = .05$ level, we also declare as significant (by Tocher's modification) a proportion (.01791) of the cases when a occurs in the sampling distribution. Whether we may consider our observed case as one of those in the proportion is determined by a table of random numbers.

Summary of procedure. These are the steps in the use of the Fisher test:

1. Cast the observed frequencies in a 2 $\times$ 2 table.

2. Determine the marginal totals. Each set of marginal totals sums to N, the number of independent cases observed.

3. The method of deciding whether or not to reject H_0 depends on whether or not exact probabilities are required:

   a. For a test of significance, refer to Table I.

   b. For an exact probability, the recursive use of formula (6.1) is required.

In either case, the value yielded will be for a one-tailed test. For a two-tailed test, the significance level shown by Table I or the p yielded by the use of formula (6.1) must be doubled.

4. If the significance level shown by Table I or the p yielded by the use of formula (6.1) is equal to or less than α, reject H_0.

5. If the observed frequencies are insignificant but all more extreme possible outcomes with the same marginal totals would be significant, use Tocher's modification to determine whether or not to reject H_0 for a one-tailed test.

Power

With Tocher's modification, the Fisher test is the most powerful of one-tailed tests (in the sense of Neyman and Pearson) for data of the kind for which the test is appropriate (Cochran, 1952).

References

Other discussions of the Fisher Test may be found in Barnard (1947), Cochran (1952), Finney (1948), Fisher (1934, sec. 21.02), McNemar (1955, pp. 240–242), and Tocher (1950).

THE χ^2 TEST FOR TWO INDEPENDENT SAMPLES

Function

When the data of research consist of frequencies in discrete categories, the χ^2 test may be used to determine the significance of differences between two independent groups. The measurement involved may be as weak as nominal scaling.

The hypothesis under test is usually that the two groups differ with respect to some characteristic and therefore with respect to the relative frequency with which group members fall in several categories. To test this hypothesis, we count the number of cases from each group which fall in the various categories, and compare the proportion of cases from one group in the various categories with the proportion of cases from the other group. For example, we might test whether two political groups differ in their agreement or disagreement with some opinion, or we might test whether the sexes differ in the frequency with which they choose certain leisure time activities, etc.

Method

The null hypothesis may be tested by

$$\chi^2 = \sum_{i=1}^{r} \sum_{j=1}^{k} \frac{(O_{ij} - E_{ij})^2}{E_{ij}} \tag{6.3}$$

where O_{ij} = observed number of cases categorized in ith row of jth column

E_{ij} = number of cases expected under H_0 to be categorized in ith row of jth column

$\sum_{i=1}^{r} \sum_{j=1}^{k}$ directs one to sum over all (r) rows and all (k) columns,

i.e., to sum over all cells

The values of χ^2 yielded by formula (6.3) are distributed approximately as chi square with $df = (r - 1)(k - 1)$, where r = the number of rows and k = the number of columns in the contingency table.

To find the expected frequency for each cell (E_{ij}), multiply the two marginal totals common to a particular cell, and then divide this product by the total number of cases, N.

We may illustrate the method of finding expected values by a simple example, using artificial data. Suppose we wished to test whether tall and short persons differ with respect to leadership qualities. Table 6.7

TABLE 6.7. HEIGHT AND LEADERSHIP
(Artificial data)

	Short	Tall	Total
Leader	12	32	44
Follower	22	14	36
Unclassifiable	9	6	15
Total	43	52	95

shows the frequencies with which 43 short people and 52 tall people are categorized as "leaders," "followers," and as "unclassifiable." Now the null hypothesis would be that height is independent of leader-follower position, i.e., that the proportion of tall people who are leaders is the same as the proportion of short people who are leaders, that the proportion of tall people who are followers is the same as the proportion of short people who are followers, etc. With such a hypothesis, we may determine the expected frequency for each cell by the method indicated.

TABLE 6.8. HEIGHT AND LEADERSHIP: OBSERVED AND EXPECTED FREQUENCIES
(Artificial data)

	Short		Tall		Total
Leader	*19.9*	12	*24.1*	32	44
Follower	*16.3*	22	*19.7*	14	36
Unclassifiable	*6.8*	9	*8.2*	6	15
Total	43		52		95

In each case we multiply the two marginal totals common to a particular cell, and then divide this product by N to obtain the expected frequency.

Thus, for example, the expected frequency for the lower right-hand cell in Table 6.7 is $E_{32} = \dfrac{(52)(15)}{95} = 8.2$. Table 6.8 shows the expected frequencies for each of the six cells for the data shown in Table 6.7. In each case the expected frequencies are shown in italics in Table 6.8, which also shows the various observed frequencies.

Now if .the observed frequencies are in close agreement with the expected frequencies, the differences $(O_{ij} - E_{ij})$ will of course be small, and consequently the value of χ^2 will be small. With a small value of χ^2 we may not reject the null hypothesis that the two sets of characteristics are independent of each other. However, if some or many of the differences are large, then the value of χ^2 will also be large. The larger is χ^2, the more likely it is that the two groups differ with respect to the classifications.

The sampling distribution of χ^2 as defined by formula (6.3) can be shown to be approximated by a chi-square[1] distribution with

$$df = (r - 1)(k - 1)$$

The probabilities associated with various values of chi square are given in Table C of the Appendix. If an observed value of χ^2 is equal to or greater than the value given in Table C for a particular level of significance, at a particular df, then H_0 may be rejected at that level of significance.

Notice that there is a different sampling distribution for every value of df. That is, the significance of any particular value of χ^2 depends on the number of degrees of freedom in the data from which it was computed. The size of df reflects the number of observations that are free to vary after certain restrictions have been placed on the data. (Degrees of freedom are discussed in Chap. 4.)

The degrees of freedom for an $r \times k$ contingency table may be found by

$$df = (r - 1)(k - 1)$$

where r = number of classifications (rows)

     k = number of groups (columns)

For the data in Table 6.8, $r = 3$ and $k = 2$, for we have 3 classifications (leader, follower, and unclassifiable) and 2 groups (tall and short). Thus the $df = (3 - 1)(2 - 1) = 2$.

---

[1] To avoid confusion, the symbol χ^2 is used for the quantity in formula (6.3) which is computed from the observed data when a χ^2 test is performed. The words "chi-square" refer to a random variable which follows the chi-square distribution, tabled in Table C.

The computation of χ^2 for the data in Table 6.8 is straightforward:

$$\chi^2 = \sum_{i=1}^{r} \sum_{j=1}^{k} \frac{(O_{ij} - E_{ij})^2}{E_{ij}} \tag{6.3}$$

$$= \frac{(12 - 19.9)^2}{19.9} + \frac{(32 - 24.1)^2}{24.1} + \frac{(22 - 16.3)^2}{16.3} + \frac{(14 - 19.7)^2}{19.7}$$

$$+ \frac{(9 - 6.8)^2}{6.8} + \frac{(6 - 8.2)^2}{8.2}$$

$$= 3.14 + 2.59 + 1.99 + 1.65 + .71 + .59$$

$$= 10.67$$

To determine the significance of $\chi^2 = 10.67$ when $df = 2$, we turn to Table C. The table shows that this value of χ^2 is significant beyond the .01 level. Therefore we could reject the null hypothesis of no differences at $\alpha = .01$.

2 × 2 contingency tables. Perhaps the most common of all uses of the χ^2 test is the test of whether an observed breakdown of frequencies in a 2 × 2 contingency table could have occurred under H_0. We are familiar with the form of such a table; an example is Table 6.1. When applying the χ^2 test to data where both r and k equal 2, formula (6.4) should be used:

$$\chi^2 = \frac{N\left(|AD - BC| - \dfrac{N}{2}\right)^2}{(A + B)(C + D)(A + C)(B + D)} \qquad df = 1 \tag{6.4}$$

This formula is somewhat easier to apply than formula (6.3), inasmuch as only one division is necessary in the computation. Moreover, it lends itself readily to machine computation. It has the additional advantage of incorporating a correction for continuity which markedly improves the approximation of the distribution of the computed χ^2 by the chi-square distribution.

Example

Adams studied the relation of vocational interests and curriculum choice to rate of withdrawal from college by bright students.[1] Her subjects were students who scored at or above the 90th percentile in college entrance tests of intelligence, and who changed their majors following matriculation. She compared those bright students whose curriculum choice was in the direction indicated as desirable by their

---

[1] Adams, Lois. 1955. A study of intellectually gifted students who withdrew from the Pennsylvania State University. Unpublished master's thesis, Pennsylvania State University.

scores on the Strong Vocational Interest Test (such a change was called "positive") with those bright students whose curriculum change was in a direction contrary to that suggested by their tested interests. Her hypothesis was that those who made positive curricular changes would more frequently remain in school.

i. *Null Hypothesis.* H_0: there is no difference between the two groups (positive curriculum changers and negative curriculum changers) in the proportion of members who remain in college. H_1: a greater proportion of students who make positive curriculum changes remain in college than is the case with those who make negative curriculum changes.

ii. *Statistical Test.* The χ^2 test for two independent samples is chosen because the two groups (positive and negative curriculum changers) are independent, and because the "scores" under study are frequencies in discrete categories (withdrew and remained).

iii. *Significance Level.* Let $\alpha = .05$. $N =$ the number of students in the sample $= 80$.

iv. *Sampling Distribution.* χ^2 as computed from formula (6.4) has a sampling distribution which is approximated by the chi-square distribution with $df = 1$. Critical values of chi square are given in Table C.

v. *Rejection Region.* The region of rejection consists of all values of χ^2 which are so large that the probability associated with their occurrence is equal to or less than $\alpha = .05$. Since H_1 predicts the direction of the difference between the two groups, the region of rejection is one-tailed. Table C shows that for a one-tailed test, when $df = 1$, a χ^2 of 2.71 or larger has probability of occurrence under H_0 of $p = \frac{1}{2}(.10) = .05$. Therefore the region of rejection consists of all $\chi^2 \geq 2.71$ if the direction of the results is that predicted by H_1.

vi. *Decision.* Adams' findings are presented in Table 6.9. This table shows that of the 56 bright students who made positive curriculum changes, 10 withdrew and 46 remained in college. Of the

TABLE 6.9. CURRICULUM CHANGE AND WITHDRAWAL FROM
COLLEGE AMONG BRIGHT STUDENTS

| | Direction of curriculum change | | Total |
	Positive	Negative	
Withdrew	10	11	21
Remained	46	13	59
Total	56	24	80

24 who made negative changes, 11 withdrew from college and 13 remained.

The value of χ^2 for these data is

$$\chi^2 = \frac{N\left(|AD - BC| - \frac{N}{2}\right)^2}{(A + B)(C + D)(A + C)(B + D)} \qquad (6.4)$$

$$= \frac{80(|(10)(13) - (11)(46)| - \frac{80}{2})^2}{(21)(59)(56)(24)}$$

$$= \frac{80(336)^2}{1,665,216}$$

$$= 5.42$$

The probability of occurrence under H_0 for $\chi^2 \geq 5.42$ with $df = 1$ is $p < \frac{1}{2}(.02) = p < .01$. Inasmuch as this p is less than $\alpha = .05$, the decision is to reject H_0 in favor of H_1. We conclude that bright students who make "positive" curriculum changes remain in college more frequently than do bright students who make "negative" curriculum changes.

Small expected frequencies. The χ^2 test is applicable to data in a contingency table only if the expected frequencies are sufficiently large. The size requirements for expected frequencies are discussed below. When the observed expected frequencies do not meet these requirements, one may increase their values by combining cells, i.e., by combining adjacent classifications and thereby reducing the number of cells. This may be properly done only if such combining does not rob the data of their meaning. In our fictitious "study" of height and leadership, of course, any combining of categories would have rendered the data useless for testing our hypothesis. The researcher may usually avoid this problem by planning in advance to collect a fairly large number of cases relative to the number of classifications he wishes to use in his analysis.

Summary of procedure. These are the steps in the use of the χ^2 test for two independent samples:

1. Cast the observed frequencies in a $k \times r$ contingency table, using the k columns for the groups and the r rows for the conditions. Thus for this test $k = 2$.

2. Determine the expected frequency for each cell by finding the product of the marginal totals common to it and dividing this by N. (N is the sum of each group of marginal totals. It represents the total number of *independent* observations. Inflated N's invalidate the test.) Step 2 is unnecessary if the data are in a 2×2 table and thus formula (6.4) is to be used.

3. For a 2 × 2 table, compute χ^2 by formula (6.4). When r is larger than 2, compute χ^2 by formula (6.3).

4. Determine the significance of the observed χ^2 by reference to Table C. For a one-tailed test, halve the significance level shown. If the probability given by Table C is equal to or smaller than α, reject H_0 in favor of H_1.

When to Use the χ^2 Test

As we have already noted, the χ^2 test requires that the expected frequencies (E_{ij}) in each cell should not be too small. When they are smaller than minimal, the test may not be properly or meaningfully used. Cochran (1954) makes these recommendations:

The 2 × 2 case. If the frequencies are in a 2 × 2 contingency table, the decision concerning the use of χ^2 should be guided by these considerations:

1. When $N > 40$, use χ^2 corrected for continuity, i.e., use formula (6.4).

2. When N is between 20 and 40, the χ^2 test [formula (6.4)] may be used *if* all expected frequencies are 5 or more. If the smallest expected frequency is less than 5, use the Fisher test (pages 94 to 104).

3. When $N < 20$, use the Fisher test in all cases.

Contingency tables with *df* **larger than 1.** When k is larger than 2 (and thus $df > 1$), the χ^2 test may be used if fewer than 20 per cent of the cells have an expected frequency of less than 5 and if no cell has an expected frequency of less than 1. If these requirements are not met by the data in the form in which they were originally collected, the researcher must combine adjacent categories in order to increase the expected frequencies in the various cells. Only after he has combined categories to meet the above requirements may he meaningfully apply the χ^2 test.

When $df > 1$, χ^2 tests are insensitive to the effects of order, and thus when a hypothesis takes order into account, χ^2 may not be the best test. The reader may consult Cochran (1954) for methods that strengthen the common χ^2 tests when H_0 is tested against specific alternatives.

Power

When the χ^2 test is used there is usually no clear alternative and thus the exact power of the test is difficult to compute. However, Cochran (1952) has shown that the limiting power distribution of χ^2 tends to 1 as N becomes large.

References

For other discussions of the χ^2 test, the reader may refer to Cochran (1952; 1954), Dixon and Massey (1951, chap. 13), Edwards (1954, chap. 18), Lewis and Burke (1949), McNemar (1955, chap. 13), and Walker and Lev (1953, chap. 4.).

THE MEDIAN TEST

Function

The median test is a procedure for testing whether two independent groups differ in central tendencies. More precisely, the median test will give information as to whether it is likely that two independent groups (not necessarily of the same size) have been drawn from populations with the same median. The null hypothesis is that the two groups are from populations with the same median; the alternative hypothesis may be that the median of one population is *different* from that of the other (two-tailed test) or that the median of one population is *higher* than that of the other (one-tailed test). The test may be used whenever the scores for the two groups are in at least an ordinal scale.

Rationale and Method

To perform the median test, we first determine the median score for the combined group (i.e., the median for all scores in both samples). Then we dichotomize both sets of scores at that combined median, and cast these data in a 2 × 2 table like Table 6.10.

TABLE 6.10. MEDIAN TEST: FORM FOR DATA

	Group I	Group II	Total
No. of scores above combined median	A	B	$A + B$
No. of scores below combined median	C	D	$C + D$
Total	$A + C$	$B + D$	$N = n_1 + n_2$

Now if both group I and group II are samples from populations whose median is the same, we would expect about half of each group's scores to be above the combined median and about half to be below. That is, we would expect frequencies A and C to be about equal, and frequencies B and D to be about equal.

It can be shown (Mood, 1950, pp. 394–395) that if A is the number of cases in group I which fall above the combined median, and if B is the number of cases in group II which fall above the combined median, then the sampling distribution of A and B under the null hypothesis (H_0 is that $A = \frac{1}{2}n_1$ and $B = \frac{1}{2}n_2$) is the hypergeometric distribution

$$p_{(A,B)} = \frac{\binom{A + C}{A}\binom{B + D}{B}}{\binom{n_1 + n_2}{A + B}} \tag{6.5}$$

Therefore if the total number of cases in both groups $(n_1 + n_2)$ is small, one may use the Fisher test (pages 96 to 104) to test H_0. If the total number of cases is sufficiently large, the χ^2 test with $df = 1$ (page 107) may be used to test H_0.

When analyzing data split at the median, the researcher should be guided by these considerations in choosing between the Fisher test and the χ^2 test:

1. When $n_1 + n_2$ is larger than 40, use χ^2 corrected for continuity, i.e., use formula (6.4).

2. When $n_1 + n_2$ is between 20 and 40 and when no cell has an expected frequency[1] of less than 5, use χ^2 corrected for continuity [formula (6.4)]. If the smallest expected frequency is less than 5, use the Fisher test.

3. When $n_1 + n_2$ is less than 20, use the Fisher test.

One difficulty may arise in the computation of the median test: several scores may fall right at the combined median. If this happens, the researcher has two alternatives: (a) if $n_1 + n_2$ is large, and if only a few cases fall at the combined median, those few cases may be dropped from the analysis, or (b) the groups may be dichotomized as those scores which *exceed* the median and those which do not. In this case, the troublesome scores would be included in the second category.

Example

In a cross-cultural test of some behavior theory hypotheses adapted from psychoanalytic theory,[2] Whiting and Child studied the relation between child-rearing practices and customs related to illness in various nonliterate cultures. One hypothesis of their study, derived from the notion of negative fixation, was that oral explanations of illness would be used in societies in which the socialization of oral drives is such as to produce anxiety. Typical oral explanations of illness are these: illness results from eating poison, illness results from drinking certain liquids, illness results from verbal spells and incantations performed by others. Judgments of the typical oral socialization anxiety in any society were based on the rapidity of oral socialization, the severity of oral socialization, the frequency of punishment typical in oral socialization, and the severity of emotional conflict typically evidenced by the children during the period of oral socialization.

Excerpts from ethnological reports of nonliterate cultures were used in the collection of the data. Using only excerpts concerning

---

[1] The method for computing expected frequencies is given on pages 105 and 106.

[2] Whiting, J. W. M., and Child, I. L. 1953. *Child training and personality.* New Haven: Yale Univer. Press.

customs relating to illness, judges classified the societies into two groups: those with oral explanations of illness present and those with oral explanations absent. Other judges, using the excerpts concerning child-rearing practices, rated each society on the degree of oral socialization anxiety typical in its children. For the 39 societies for which judgments of the presence or absence of oral explanations were possible, these ratings ranged from 6 to 17.

i. *Null Hypothesis.* H_0: there is no difference between the median oral socialization anxiety in societies which give oral explanations of illness and the median oral socialization anxiety in societies which do not give oral explanations of illness. H_1: the median oral socialization anxiety in societies with oral explanations present is higher than the median in societies with oral explanations absent.

ii. *Statistical Test.* The ratings constitute ordinal measures at best; thus a nonparametric test is appropriate. This choice also eliminates the necessity of assuming that oral socialization anxiety is normally distributed among the cultures sampled, as well as eliminating the necessity of assuming that the variances of the two groups sampled are equal. For the data from the two independent groups of societies, the median test may be used to test H_0.

iii. *Significance Level.* Let $\alpha = .01$. $N = 39 =$ the number of societies for which ethnological information on both variables was available. $n_1 = 16 =$ the number of societies with oral explanations absent; $n_2 = 23 =$ the number of societies with oral explanations present.

iv. *Sampling Distribution.* Since we cannot at this time state which test (Fisher test or χ^2 test) will be used for the scores split at the median, since $n_1 + n_2 = 39$ is between 20 and 40 and therefore our choice must be determined by the size of the smallest expected frequency, we cannot state the sampling distribution.

v. *Rejection Region.* Since H_1 predicts the direction of the difference, the region of rejection is one-tailed. It consists of all outcomes in a median-split table which are in the predicted direction and which are so extreme that the probability associated with their occurrence under H_0 (as determined by the appropriate test) is equal to or less than $\alpha = .01$.

vi. *Decision.* Table 6.11 shows the ratings assigned to each of the 39 societies. These are divided at the combined median for the $n_1 + n_2$ ratings. (We have followed Whiting and Child in calling 10.5 the median of the 39 ratings.) Table 6.12 shows these data cast in the form for the median test. Since none of the expected frequencies is less than 5, and since $n_1 + n_2 > 20$, we may use the χ^2 test to test H_0:

$$\chi^2 = \frac{N\left(|AD - BC| - \dfrac{N}{2}\right)^2}{(A + B)(C + D)(A + C)(B + D)} \tag{6.4}$$

$$= \frac{39(|(3)(6) - (17)(13)| - \frac{39}{2})^2}{(20)(19)(16)(23)}$$

$$= 9.39$$

TABLE 6.11. ORAL SOCIALIZATION ANXIETY AND ORAL EXPLANATIONS
OF ILLNESS*
(The name of each society is preceded by its rating on
oral socialization anxiety)

	Societies with oral explanations absent	Societies with oral explanations present
Societies above median on oral socialization anxiety	13 Lapp 12 Chamorro 12 Samoans	17 Marquesans 16 Dobuans 15 Baiga 15 Kwoma 15 Thonga 14 Alorese 14 Chagga 14 Navaho 13 Dahomeans 13 Lesu 13 Masai 12 Lepcha 12 Maori 12 Pukapukans 12 Trobrianders 11 Kwakiutl 11 Manus
Societies below median on oral socialization anxiety	10 Arapesh 10 Balinese 10 Hopi 10 Tanala 9 Paiute 8 Chenchu 8 Teton 7 Flathead 7 Papago 7 Venda 7 Warrau 7 Wogeo 6 Ontong-Javanese	10 Chiricahua 10 Comanche 10 Siriono 8 Bena 8 Slave 6 Kurtatchi

* Reproduced from Table 4 of Whiting, J. W. M., and Child, I. L. 1953. *Child training and personality*. New Haven: Yale Univer. Press, p. 156, with the kind permission of the authors and the publisher.

TABLE 6.12. ORAL SOCIALIZATION ANXIETY AND ORAL EXPLANATIONS
OF ILLNESS

	Societies with oral explanations absent	Societies with oral explanations present	Total
Societies above median on oral socialization anxiety	3	17	20
Societies below median on oral socialization anxiety	13	6	19
Total	16	23	39

Reference to Table C shows that $\chi^2 \geq 9.39$ with $df = 1$ has probability of occurrence under H_0 of $p < \frac{1}{2}(.01) = p < .005$ for a one-tailed test. Thus our decision is to reject H_0 for $\alpha = .01$.[1] We conclude that the median oral socialization anxiety is higher in societies with oral explanations of illness present than is the median in societies with oral explanations absent.

Summary of procedure. These are the steps in the use of the median test:

1. Determine the combined median of the $n_1 + n_2$ scores.

2. Split each group's scores at that combined median. Enter the resultant frequencies in a table like Table 6.10. If many scores fall at the combined median, split the scores into these categories: those which exceed the median and those which do not.

3. Find the probability of the observed values by either the Fisher test or the χ^2 test, choosing between these according to the criteria given above.

4. If the p yielded by that test is equal to or smaller than α, reject H_0.

Power-Efficiency

Mood (1954) has shown that when the median test is applied to data measured in at least an interval scale from normal distributions with common variance (i.e., data that might properly be analyzed by the parametric t test), it has the same power-efficiency as the sign test. That is, its power-efficiency is about 95 per cent for $n_1 + n_2$ as low as 6. This power-efficiency decreases as the sample sizes increase, reaching an eventual asymptotic efficiency of $2/\pi = 63$ per cent.

[1] This decision agrees with that reached by Whiting and Child. Using the parametric t test on these data, they found that $t = 4.05$, $p < .0005$.

References

Discussions of the median test are contained in Brown and Mood (1951), Mood (1950, pp. 394–395), and Moses (1952a).

THE MANN-WHITNEY U TEST

Function

When at least ordinal measurement has been achieved, the Mann-Whitney U test may be used to test whether two independent groups have been drawn from the same population. This is one of the most powerful of the nonparametric tests, and it is a most useful alternative to the parametric t test when the researcher wishes to avoid the t test's assumptions, or when the measurement in the research is weaker than interval scaling.

Suppose we have samples from two populations, population A and population B. The null hypothesis is that A and B have the same distribution. The alternative hypothesis, H_1, against which we test H_0, is that A is stochastically larger than B, a directional hypothesis. We may accept H_1 if the probability that a score from A is larger than a score from B is greater than one-half. That is, if a is one observation from population A, and b is one observation from population B, then H_1 is that $p(a > b) > \frac{1}{2}$. If the evidence supports H_1, this implies that the "bulk" of population A is higher than the bulk of population B.

Of course, we might predict instead that B is stochastically larger than A. Then H_1 would be that $p(a > b) < \frac{1}{2}$. Confirmation of this assertion would imply that the bulk of B is higher than the bulk of A.

For a two-tailed test, i.e., for a prediction of differences which does not state direction, H_1 would be that $p(a > b) \neq \frac{1}{2}$.

Method

Let $n_1 =$ the number of cases in the smaller of two independent groups, and $n_2 =$ the number of cases in the larger. To apply the U test, we first combine the observations or scores from both groups, and rank these in order of increasing size. In this ranking, algebraic size is considered, i.e., the lowest ranks are assigned to the largest negative numbers, if any.

Now focus on one of the groups, say the group with n_1 cases. The value of U (the statistic used in this test) is given by the number of times that a score in the group with n_2 cases precedes a score in the group with n_1 cases in the ranking.

For example, suppose we had an experimental group of 3 cases and a control group of 4 cases. Here $n_1 = 3$ and $n_2 = 4$. Suppose these were the scores:

E scores	9	11	15	
C scores	6	8	10	13

To find U, we first rank these scores in order of increasing size, being careful to retain each score's identity as either an E or C score:

6	8	9	10	11	13	15
C	C	E	C	E	C	E

Now consider the control group, and count the number of E scores that precede each score in the control group. For the C score of 6, no E score precedes. This is also true for the C score of 8. For the next C score (10), one E score precedes. And for the final C score (13), two E scores precede. Thus $U = 0 + 0 + 1 + 2 = 3$. The number of times that an E score precedes a C score is $3 = U$.

The sampling distribution of U under H_0 is known, and with this knowledge we can determine the probability associated with the occurrence under H_0 of any U as extreme as an observed value of U.

Very small samples. When neither n_1 nor n_2 is larger than 8, Table J of the Appendix may be used to determine the exact probability associated with the occurrence under H_0 of any U as extreme as an observed value of U. The reader will observe that Table J is made up of six separate subtables, one for each value of n_2, from $n_2 = 3$ to $n_2 = 8$. To determine the probability under H_0 associated with his data, the researcher need know only n_1 (the size of the smaller group), n_2, and U. With this information he may read the value of p from the subtable appropriate to his value of n_2.

In our example, $n_1 = 3$, $n_2 = 4$, and $U = 3$. The subtable for $n_2 = 4$ in Table J shows that $U \leq 3$ has probability of occurrence under H_0 of $p = .200$.

The probabilities given in Table J are one-tailed. For a two-tailed test, the value of p given in the table should be doubled.

Now it may happen that the observed value of U is so large that it does not appear in the subtable for the observed value of n_2. Such a value arises when the researcher focuses on the "wrong" group in determining U. We shall call such a too-large value U'. For example, suppose that in the above case we had counted the number of C scores preceding each E score rather than counting the number of E scores preceding each C score. We would have found that $U = 2 + 3 + 4 = 9$. The subtable for $n_2 = 4$ does not go up to $U = 9$. We therefore denote our observed value as $U' = 9$. We can transform any U' to U by

$$U = n_1 n_2 - U' \qquad (6.6)*$$

* $p(U \geq U') = p(U \leq n_1 n_2 - U')$.

In our example, by this transformation $U = (3)(4) - 9 = 3$. Of course this is the U we found directly when we counted the number of E scores preceding each C score.

Example for Very Small Samples

Solomon and Coles[1] studied whether rats would generalize learned imitation when placed under a new drive and in a new situation. Five rats were trained to imitate leader rats in a T maze. They were trained to follow the leaders when hungry, in order to attain a food incentive. Then the 5 rats were each transferred to a shock-avoidance situation, where imitation of leader rats would have enabled them to avoid electric shock. Their behavior in the shock-avoidance situation was compared to that of 4 controls who had had no previous training to follow leaders. The hypothesis was that the 5 rats who had already been trained to imitate would transfer this training to the new situation, and thus would reach the learning criterion in the shock-avoidance situation sooner than would the 4 control rats. The comparison is in terms of how many trials each rat took to reach a criterion of 10 correct responses in 10 trials.

i. *Null Hypothesis.* H_0: the number of trials to the criterion in the shock-avoidance situation is the same for rats previously trained to follow a leader to a food incentive as for rats not previously trained. H_1: rats previously trained to follow a leader to a food incentive will reach the criterion in the shock-avoidance situation in fewer trials than will rats not previously trained.

ii. *Statistical Test.* The Mann-Whitney U test is chosen because this study employs two independent samples, uses small samples, and uses measurement (number of trials to criterion as an index to speed of learning) which is probably at most in an ordinal scale.

iii. *Significance Level.* Let $\alpha = .05$. $n_1 = 4$ control rats, and $n_2 = 5$ experimental rats.

iv. *Sampling Distribution.* The probabilities associated with the occurrence under H_0 of values as small as an observed U for n_1, $n_2 \leq 8$ are given in Table J.

v. *Rejection Region.* Since H_1 states the direction of the predicted difference, the region of rejection is one-tailed. It consists of all values of U which are so small that the probability associated with their occurrence under H_0 is equal to or less than $\alpha = .05$.

vi. *Decision.* The number of trials to criterion required by the E

---

[1] Solomon, R. L., and Coles, M. R. 1954. A case of failure of generalization of imitation across drives and across situations. *J. Abnorm. Soc. Psychol.*, **49**, 7–13. Only two of the groups studied by these investigators are included in this example.

and C rats were:

E rats	78	64	75	45	82
C rats	110	70	53	51	

We arrange these scores in the order of their size, retaining the identity of each:

45	51	53	64	70	75	78	82	110
E	C	C	E	C	E	E	E	C

We obtain U by counting the number of E scores preceding each C score: $U = 1 + 1 + 2 + 5 = 9$.

In Table J, we locate the subtable for $n_2 = 5$. We see that $U \leq 9$ when $n_1 = 4$ has a probability of occurrence under H_0 of $p = .452$. Our decision is that the data do not give evidence which justify rejecting H_0 at the previously set level of significance. The conclusion is that these data do not support the hypothesis that previous training to imitate will generalize across situations and across drives.[1]

n_2 **between 9 and 20.** If n_2 (the size of the larger of the two independent samples) is larger than 8, Table J may not be used. When n_2 is between 9 and 20, significance tests may be made with the Mann-Whitney test by using Table K of the Appendix which gives critical values of U for significance levels .001, .01, .025, and .05 for a one-tailed test. For a two-tailed test, the significance levels given are .002, .02, .05, and .10.

Notice that this set of tables gives critical values of U, and does not give exact probabilities (as does Table J). That is, if an observed U for a particular $n_1 \leq 20$ and n_2 between 9 and 20 is equal to or less than that value given in the table, H_0 may be rejected at the level of significance indicated at the head of that table.

For example, if $n_1 = 6$ and $n_2 = 13$, a U of 12 enables us to reject H_0 at $\alpha = .01$ for a one-tailed test, and to reject H_0 at $\alpha = .02$ for a two-tailed test.

Computing the value of U. For fairly large values of n_1 and n_2, the counting method of determining the value of U may be rather tedious. An alternative method, which gives identical results, is to assign the

---

[1] Solomon and Coles report the same conclusion. The statistical test which they utilized is not disclosed.

rank of 1 to the lowest score in the combined $(n_1 + n_2)$ group of scores, assign rank 2 to the next lowest score, etc.   Then

$$U = n_1n_2 + \frac{n_1(n_1 + 1)}{2} - R_1 \qquad (6.7a)$$

or, equivalently,

$$U = n_1n_2 + \frac{n_2(n_2 + 1)}{2} - R_2 \qquad (6.7b)$$

where R_1 = sum of the ranks assigned to group whose sample size is n_1
R_2 = sum of the ranks assigned to group whose sample size is n_2

For example, we might have used this method in finding the value of U for the data given in the example for small samples above.   The E and C scores for that example are given again in Table 6.13, with their ranks.

TABLE 6.13. TRIALS TO CRITERION OF E AND C RATS

E Score	Rank	C Score	Rank
78	7	110	9
64	4	70	5
75	6	53	3
45	1	51	2
82	8		
	$R_2 = 26$		$R_1 = 19$

For those data, $R_1 = 19$ and $R_2 = 26$, and it will be remembered that $n_1 = 4$ and $n_2 = 5$.   By applying formula (6.7b), we have

$$U = (4)(5) + \frac{5(5 + 1)}{2} - 26$$
$$= 9$$

$U = 9$ is of course exactly the value we found earlier by counting.

Formulas (6.7a) and (6.7b) yield different U's.   It is the smaller of these that we want.   The larger value is U'.   The investigator should check whether he has found U' rather than U by applying the transformation

$$U = n_1n_2 - U' \qquad (6.6)$$

The smaller of the two values, U, is the one whose sampling distribution is the basis for Table K.   Although this value can be found by computing both formulas (6.7a) and (6.7b) and choosing the smaller of the two results, a simpler method is to use only one of those formulas and then find the other value by formula (6.6).

Large samples (n_2 **larger than 20**).   Neither Table J nor Table K is usable when $n_2 > 20$.   However, it has been shown (Mann and Whitney,

1947) that as n_1, n_2 increase in size, the sampling distribution of U rapidly approaches the normal distribution, with

$$\text{Mean} = \mu_U = \frac{n_1 n_2}{2}$$

and Standard deviation $= \sigma_U = \sqrt{\dfrac{(n_1)(n_2)(n_1 + n_2 + 1)}{12}}$

That is, when $n_2 > 20$ we may determine the significance of an observed value of U by

$$z = \frac{U - \mu_U}{\sigma_U} = \frac{U - \dfrac{n_1 n_2}{2}}{\sqrt{\dfrac{(n_1)(n_2)(n_1 + n_2 + 1)}{12}}} \tag{6.8}$$

which is practically normally distributed with zero mean and unit variance. That is, the probability associated with the occurrence under H_0 of values as extreme as an observed z may be determined by reference to Table A of the Appendix.

When the normal approximation to the sampling distribution of U is used in a test of H_0, it does not matter whether formula (6.7a) or (6.7b) is used in the computation of U, for the absolute value of z yielded by formula (6.8) will be the same if either is used. The sign of the z depends on whether U or U' was used, but the value does not.

Example for Large Samples

For our example, we will reexamine the Whiting and Child data which we have already analyzed by the median test (on pages 112 to 115).

i. *Null Hypothesis.* H_0: oral socialization anxiety is equally severe in both societies with oral explanations of illness present and societies with oral explanations absent. H_1: societies with oral explanations of illness present are (stochastically) higher in oral socialization anxiety than societies which do not have oral explanations of illness.

ii. *Statistical Test.* The two groups of societies constitute two independent groups, and the measure of oral socialization anxiety (rating scale) constitutes an ordinal measure at best. For these reasons the Mann-Whitney U test is appropriate for analyzing these data.

iii. *Significance Level.* Let $\alpha = .01$. $n_1 = 16 =$ the number of societies with oral explanations absent; $n_2 = 23 =$ the number of societies with oral explanations present.

iv. *Sampling Distribution.* For $n_2 > 20$, formula (6.8) yields values of z. The probability associated with the occurrence under H_0 of values as extreme as an observed z may be determined by reference to Table A.

v. *Rejection Region.* Since H_1 predicts the direction of the difference, the region of rejection is one-tailed. It consists of all values of z (from data in which the difference is in the predicted direction) which are so extreme that their associated probability under H_0 is equal to or less than $\alpha = .01$.

vi. *Decision.* The ratings assigned to each of the 39 societies are shown in Table 6.14, together with the rank of each in the combined

TABLE 6.14. ORAL SOCIALIZATION ANXIETY AND ORAL EXPLANATIONS OF ILLNESS

Societies with oral explanations absent	Rating on oral socialization anxiety	Rank	Societies with oral explanations present	Rating on oral socialization anxiety	Rank
Lapp	13	29.5	Marquesans	17	39
Chamorro	12	24.5	Dobuans	16	38
Samoans	12	24.5	Baiga	15	36
Arapesh	10	16	Kwoma	15	36
Balinese	10	16	Thonga	15	36
Hopi	10	16	Alorese	14	33
Tanala	10	16	Chagga	14	33
Paiute	9	12	Navaho	14	33
Chenchu	8	9.5	Dahomeans	13	29.5
Teton	8	9.5	Lesu	13	29.5
Flathead	7	5	Masai	13	29.5
Papago	7	5	Lepcha	12	24.5
Venda	7	5	Maori	12	24.5
Warrau	7	5	Pukapukans	12	24.5
Wogeo	7	5	Trobrianders	12	24.5
Ontong-Javanese	6	1.5	Kwakiutl	11	20.5
			Manus	11	20.5
			Chiricahua	10	16
	$R_1 = 200.0$		Comanche	10	16
			Siriono	10	16
			Bena	8	9.5
			Slave	8	9.5
			Kurtatchi	6	1.5
					$R_2 = 580.0$

group. Notice that tied ratings are assigned the average of the tied ranks. For these data, $R_1 = 200.0$ and $R_2 = 580.0$. The value of U may be found by substituting the observed values in formula (6.7a):

$$U = n_1 n_2 + \frac{n_1(n_1 + 1)}{2} - R_1 \qquad (6.7a)$$

$$= (16)(23) + \frac{16(16 + 1)}{2} - 200$$

$$= 304$$

Knowing that $U = 304$, we may find the value of z by substituting in formula (6.8):

$$z = \frac{U - \dfrac{n_1 n_2}{2}}{\sqrt{\dfrac{(n_1)(n_2)(n_1 + n_2 + 1)}{12}}} \qquad (6.8)$$

$$= \frac{304 - \dfrac{(16)(23)}{2}}{\sqrt{\dfrac{(16)(23)(16 + 23 + 1)}{12}}}$$

$$= 3.43$$

Reference to Table A reveals that $z \geq 3.43$ has a one-tailed probability under H_0 of $p < .0003$. Since this p is smaller than $\alpha = .01$, our decision is to reject H_0 in favor of H_1.* We conclude that societies with oral explanations of illness present are (stochastically) higher in oral socialization anxiety than societies with oral explanations absent.

It is important to notice that for these data the Mann-Whitney U test exhibits greater power to reject H_0 than the median test. Testing a similar hypothesis about these data, the median test yielded a value which permitted rejection of H_0 at the $p < .005$ level (one-tailed test), whereas the Mann-Whitney test yielded a value which permitted rejection of H_0 at the $p < .0003$ level (one-tailed test). The fact that the Mann-Whitney test is more powerful than the median test is not surprising, inasmuch as it considers the rank value of each observation rather than simply its location with respect to the combined median, and thus uses more of the information in the data.

Ties. The Mann-Whitney test assumes that the scores represent a distribution which has underlying continuity. With very precise measurement of a variable which has underlying continuity, the probability of a tie is zero. However, with the relatively crude measures which we typically employ in behavioral scientific research, ties may well occur.

* As we have already noted, Whiting and Child reached the same decision on the basis of the parametric t test. They found that $t = 4.05$, $p < .0005$.

We assume that the two observations which obtain tied scores are really different, but that this difference is simply too refined or minute for detection by our crude measures.

When tied scores occur, we give each of the tied observations the average of the ranks they would have had if no ties had occurred.

If the ties occur between two or more observations in the same group, the value of U is not affected. But if ties occur between two or more observations involving both groups, the value of U is affected. Although the effect is usually negligible, a correction for ties is available for use with the normal curve approximation which we employ for large samples.

The effect of tied ranks is to change the variability of the set of ranks. Thus the correction for ties must be applied to the standard deviation of the sampling distribution of U. Corrected for ties, the standard deviation becomes

$$\sigma_U = \sqrt{\left(\frac{n_1 n_2}{N(N-1)}\right)\left(\frac{N^3 - N}{12} - \Sigma T\right)}$$

where $N = n_1 + n_2$

$T = \dfrac{t^3 - t}{12}$ (where t is the number of observations tied for a given rank)

ΣT is found by summing the T's over all groups of tied observations

With the correction for ties, we find z by

$$z = \frac{U - \dfrac{n_1 n_2}{2}}{\sqrt{\left(\dfrac{n_1 n_2}{N(N-1)}\right)\left(\dfrac{N^3 - N}{12} - \Sigma T\right)}} \tag{6.9}$$

It may be seen that if there are no ties, the above expression reduces directly to that given originally for z [formula (6.8)].

The use of the correction for ties may be illustrated by applying that correction to the data in Table 6.14. For those data,

$$n_1 + n_2 = 16 + 23 = 39 = N$$

We observe these tied groups:

2 scores of 6
5 scores of 7
4 scores of 8
7 scores of 10
2 scores of 11
6 scores of 12
4 scores of 13
3 scores of 14
3 scores of 15

Thus we have t's of 2, 5, 4, 7, 2, 6, 4, 3, and 3. To find ΣT, we sum the values of $\dfrac{t^3 - t}{12}$ for each of these tied groups:

$$\Sigma T = \frac{2^3 - 2}{12} + \frac{5^3 - 5}{12} + \frac{4^3 - 4}{12} + \frac{7^3 - 7}{12} + \frac{2^3 - 2}{12} + \frac{6^3 - 6}{12}$$
$$+ \frac{4^3 - 4}{12} + \frac{3^3 - 3}{12} + \frac{3^3 - 3}{12}$$

$$= .5 + 10.0 + 5.0 + 28.0 + .5 + 17.5 + 5.0 + 2.0 + 2.0$$
$$= 70.5$$

Thus for the data in Table 6.14, $n_1 = 16$, $n_2 = 23$, $N = 39$, $U = 304$, and $\Sigma T = 70.5$. Substituting these values in formula (6.9), we have

$$z = \frac{U - \dfrac{n_1 n_2}{2}}{\sqrt{\left(\dfrac{n_1 n_2}{N(N-1)}\right)\left(\dfrac{N^3 - N}{12} - \Sigma T\right)}} \qquad (6.9)$$

$$= \frac{304 - \dfrac{(16)(23)}{2}}{\sqrt{\left(\dfrac{(16)(23)}{39(39-1)}\right)\left(\dfrac{(39)^3 - 39}{12} - 70.5\right)}}$$

$$= 3.45$$

The value of z when corrected for ties is a little larger than that found earlier when the correction was not incorporated. The difference between $z \geq 3.43$ and $z \geq 3.45$, however, is negligible in so far as the probability given by Table A is concerned. Both z's are read as having an associated probability of $p < .0003$ (one-tailed test).

As this example demonstrates, ties have only a slight effect. Even when a large proportion of the scores are tied (this example had over 90 per cent of its observations involved in ties) the effect is practically negligible. Observe, however, that the magnitude of the correction factor, ΣT, depends importantly on the *length* of the various ties, i.e., on the size of the various t's. Thus a tie of length 4 contributes 5.0 to ΣT in this example, whereas two ties of length 2 contribute together only 1.0 (that is, .5 + .5) to ΣT. And a tie of length 6 contributes 17.5, whereas two of length 3 contribute together only 2.0 + 2.0 = 4.0.

When the correction is employed, it tends to *increase* the value of z slightly, making it more significant. Therefore when we do not correct for ties our test is "conservative" in that the value of p will be slightly inflated. That is, the value of the probability associated with the observed data under H_0 will be slightly larger than that which would be found were the correction employed. The writer's recommendation is that one should correct for ties only if the proportion of ties is quite large, if some of the t's are large, or if the p which is obtained without the correction is very close to one's previously set value of α.

Summary of procedure.   These are the steps in the use of the Mann-Whitney U test:

1. Determine the values of n_1 and n_2. n_1 = the number of cases in the smaller group; n_2 = the number of cases in the larger group.

2. Rank together the scores for both groups, assigning the rank of 1 to the score which is algebraically lowest.   Ranks range from 1 to $N = n_1 + n_2$.   Assign tied observations the average of the tied ranks.

3. Determine the value of U either by the counting method or by applying formula (6.7a) or (6.7b).

4. The method for determining the significance of the observed value of U depends on the size of n_2:

   a. If n_2 is 8 or less, the exact probability associated with a value as small as the observed value of U is shown in Table J.   For a two-tailed test, double the value of p shown in that table.   If your observed U is not shown in Table J, it is U' and should be transformed to U by formula (6.6).

   b. If n_2 is between 9 and 20, the significance of any observed value of U may be determined by reference to Table K.   If your observed value of U is larger than $n_1 n_2/2$, it is U'; apply formula (6.6) for a transformation.

   c. If n_2 is larger than 20, the probability associated with a value as extreme as the observed value of U may be determined by computing the value of z as given by formula (6.8), and testing this value by referring to Table A.   For a two-tailed test, double the p shown in that table.   If the proportion of ties is very large or if the obtained p is very close to α, apply the correction for ties, i.e., use formula (6.9) rather than (6.8).

5. If the observed value of U has an associated probability equal to or less than α, reject H_0 in favor of H_1.

Power-Efficiency

If the Mann-Whitney test is applied to data which might properly be analyzed by the most powerful parametric test, the t test, its power-efficiency approaches $3/\pi = 95.5$ per cent as N increases (Mood, 1954), and is close to 95 per cent even for moderate-sized samples.   It is therefore an excellent alternative to the t test, and of course it does not have the restrictive assumptions and requirements associated with the t test.

Whitney (1948, pp. 51–56) gives examples of distributions for which the U test is superior to its parametric alternative, i.e., for which the U test has *greater* power to reject H_0.

References

For discussions of the Mann-Whitney test,[1] the reader may refer to Auble (1953), Mann and Whitney (1947), Whitney (1948), and Wilcoxon (1945).

THE KOLMOGOROV-SMIRNOV TWO-SAMPLE TEST

Function and Rationale

The Kolmogorov-Smirnov two-sample test is a test of whether two independent samples have been drawn from the same population (or from populations with the same distribution). The two-tailed test is sensitive to any kind of difference in the distributions from which the two samples were drawn—differences in location (central tendency), in dispersion, in skewness, etc. The one-tailed test is used to decide whether or not the values of the population from which one of the samples was drawn are stochastically larger than the values of the population from which the other sample was drawn, e.g., to test the prediction that the scores of an experimental group will be "better" than those of the control group.

Like the Kolmogorov-Smirnov one-sample test (pages 47 to 52), this two-sample test is concerned with the agreement between two cumulative distributions. The one-sample test is concerned with the agreement between the distribution of a set of sample values and some specified theoretical distribution. The two-sample test is concerned with the agreement between two sets of sample values.

If the two samples have in fact been drawn from the same population distribution, then the cumulative distributions of both samples may be expected to be fairly close to each other, inasmuch as they both should show only random deviations from the population distribution. If the

---

[1] Two nonparametric statistical tests which are essentially equivalent to the Mann-Whitney U test have been reported in the literature and should be mentioned here. The first of these is due to Festinger (1946). He gives a method for calculating exact probabilities and gives a two-tailed table for the .05 and .01 levels of significance for $n_1 + n_2 \leq 40$, when $n_1 \leq 12$. In addition, for n_1 from 13 to 15, values are given up to $n_1 + n_2 = 30$.

The second test is due to White (1952), who gives a method essentially the same as the Mann-Whitney test except that rather than U it employs R (the sum of the ranks of one of the groups) as its statistic. White offers two-tailed tables for the .05, .01, and .001 levels of significance for $n_1 + n_2 \leq 30$.

Inasmuch as these tests are linearly related to the Mann-Whitney test (and therefore will yield the same results in the test of H_0 for any given batch of data), it was felt that inclusion of complete discussions of them in this text would introduce unnecessary redundancy.

two sample cumulative distributions are "too far apart" at any point, this suggests that the samples come from different populations.   Thus a large enough deviation between the two sample cumulative distributions is evidence for rejecting H_0.

Method

To apply the Kolmogorov-Smirnov two-sample test, we make a cumu-- lative frequency distribution for each sample of observations, using the same intervals for both distributions.   For each interval, then, we sub- tract one step function from the other.   The test focuses on the *largest* of these observed deviations.

Let $S_{n_1}(X)$ = the observed cumulative step function of one of the samples, that is, $S_{n_1}(X) = K/n_1$, where K = the number of scores equal to or less than X.   And let $S_{n_2}(X)$ = the observed cumulative step function of the other sample, that is, $S_{n_2}(X) = K/n_2$.   Now the Kolmo- gorov-Smirnov two-sample test focuses on

$$D = \text{maximum } [S_{n_1}(X) - S_{n_2}(X)] \qquad (6.10a)$$

for a one-tailed test, and on

$$D = \text{maximum } |S_{n_1}(X) - S_{n_2}(X)| \qquad (6.10b)$$

for a two-tailed test.   The sampling distribution of D is known (Smirnov, 1948; Massey, 1951) and the probabilities associated with the occurrence of values as large as an observed D under the null hypothesis (that the two samples have come from the same distribution) have been tabled.

Notice that for a one-tailed test we find the maximum value of D *in the predicted direction* [by formula (6.10a)] and that for a two-tailed test we find the maximum *absolute* value of D [by formula (6.10b)], i.e., we find the maximum deviation irrespective of direction.   This is because in the one-tailed test, H_1 is that the population values from which one of the samples was drawn are stochastically larger than the population values from which the other sample was drawn, whereas in the two-tailed test, H_1 is simply that the two samples are from different populations.

In the use of the Kolmogorov-Smirnov test on data for which the size and number of the intervals are arbitrary, it is well to use as many intervals as are feasible.   When too few intervals are used, informa- tion may be wasted.   That is, the maximum vertical deviation D of the two cumulative step functions may be obscured by casting the data into too few intervals.

For instance, in the example presented below for the case of small samples, only 8 intervals were used, in order to simplify the exposition. As it happens, 8 intervals were sufficient, in this case, to yield a D which enabled us to reject H_0 at the predetermined level of significance.   If it had happened that with these 8 intervals the observed D had not been large enough to permit us to reject H_0, before we could accept H_0 it would be necessary for us to increase the number of intervals, in order to ascertain whether the maximum deviation D had been obscured by the use of too few intervals. It is well then to use as many intervals as are feasible to start with, so as not to waste the information inherent in the data.

Small samples. When $n_1 = n_2$, and when both n_1 and n_2 are 40 or less, Table L of the Appendix may be used in the test of the null hypothesis. The body of this table gives various values of K_D, which is defined as the numerator of the largest difference between the two cumulative distributions, i.e., the numerator of D. To read Table L, one must know the value of N (which in this case is the value of $n_1 = n_2$) and the value of K_D. Observe also whether H_1 calls for a one-tailed or a two-tailed test. With this information, one may determine the significance of the observed data.

For example, in a one-tailed test where $N = 14$, if $K_D \geq 8$ we can reject the null hypothesis at the $\alpha = .01$ level.

Example for Small Samples

Lepley[1] compared the serial learning of 10 seventh-grade pupils with the serial learning of 10 eleventh-grade pupils. His hypothesis was that the primacy effect should be less prominent in the learning of the younger subjects. The primacy effect is the tendency for the material learned early in a series to be remembered more efficiently than the material learned later in the series. He tested this hypothesis by comparing the percentage of errors made by the two groups in the first half of the series of learned material, predicting that the older group (the eleventh graders) would make relatively fewer errors in repeating the first half of the series than would the younger group.

i. *Null Hypothesis.* H_0: there is no difference in the proportion of errors made in recalling the first half of a learned series between eleventh-grade subjects and seventh-grade subjects. H_1: eleventh-graders make proportionally fewer errors than seventh-graders in recalling the first half of a learned series.

ii. *Statistical Test.* Since two small independent samples of equal size are being compared, the Kolmogorov-Smirnov two-sample test may be applied to the data.

iii. *Significance Level.* Let $\alpha = .01$. $n_1 = n_2 = N =$ the number of subjects in each group $= 10$.

[1] Lepley, W. M. 1934. Serial reactions considered as conditioned reactions. *Psychol. Monogr.*, **46**, No. 205.

iv. *Sampling Distribution.* Table L gives critical values of K_D for $n_1 = n_2$ when n_1 and n_2 are less than 40.

v. *Region of Rejection.* Since H_1 predicts the direction of the difference, the region of rejection is one-tailed. H_0 will be rejected if the value of K_D for the largest deviation in the predicted direction is so large that the probability associated with its occurrence under H_0 is equal to or less than $\alpha = .01$.

vi. *Decision.* Table 6.15 gives the percentage of each subject's

TABLE 6.15. PERCENTAGE OF TOTAL ERRORS IN FIRST HALF OF SERIES

Seventh-grade subjects	*Eleventh-grade subjects*
39.1	35.2
41.2	39.2
45.2	40.9
46.2	38.1
48.4	34.4
48.7	29.1
55.0	41.8
40.6	24.3
52.1	32.4
47.2	32.6

errors which were committed in the recall of the first half of the serially learned material. For analysis by the Kolmogorov-Smirnov test, these data were cast in two cumulative frequency distributions, shown in Table 6.16. Here $n_1 = 10$ eleventh-graders, and $n_2 = 10$ seventh-graders.

TABLE 6.16. DATA IN TABLE 6.15 CAST FOR KOLMOGOROV-SMIRNOV TEST

	Per cent of total errors in first half of series							
	24–27	28–31	32–35	36–39	40–43	44–47	48–51	52–55
$S_{10_1}(X)$	$\frac{1}{10}$	$\frac{2}{10}$	$\frac{5}{10}$	$\frac{7}{10}$	$\frac{10}{10}$	$\frac{10}{10}$	$\frac{10}{10}$	$\frac{10}{10}$
$S_{10_2}(X)$	$\frac{0}{10}$	$\frac{0}{10}$	$\frac{0}{10}$	$\frac{0}{10}$	$\frac{3}{10}$	$\frac{5}{10}$	$\frac{8}{10}$	$\frac{10}{10}$
$S_{n_1}(X) - S_{n_2}(X)$	$\frac{1}{10}$	$\frac{2}{10}$	$\frac{5}{10}$	$\frac{7}{10}$	$\frac{7}{10}$	$\frac{5}{10}$	$\frac{2}{10}$	0

Observe that the largest discrepancy between the two series is $\frac{7}{10}$. $K_D = 7$, the numerator of this largest difference. Reference to Table L reveals that when $N = 10$, a value of $K_D = 7$ is significant at the $\alpha = .01$ level for a one-tailed test. Inasmuch as the probability associated with the occurrence of a value as large as the observed value of K_D under H_0 is at most equal to the previously set level of significance, our decision is to reject H_0 in favor of H_1.* We con-

---

* Using a parametric technique, Lepley reached the same decision. He used the critical ratio technique, and rejected H_0 at $\alpha = .01$.

clude that eleventh-graders make proportionally fewer errors than seventh-graders in recalling the first half of a learned series.

Large samples : two-tailed test. When both n_1 and n_2 are larger than 40, Table M of the Appendix may be used for the Kolmogorov-Smirnov two-sample test. When this table is used, it is *not* necessary that $n_1 = n_2$.

To use this table, determine the value of D for the observed data, using formula (6.10b). Then compare that observed value with the critical one which is obtained by entering the observed values of n_1 and n_2 in the expression given in Table M. If the observed D is equal to or larger than that computed from the expression in the table, H_0 may be rejected at the level of significance (two-tailed) associated with that expression.

For example, suppose $n_1 = 55$ and $n_2 = 60$, and that a researcher wishes to make a two-tailed test at $\alpha = .05$. In the row in Table M for $\alpha = .05$, he finds the value of D which his observation must equal or exceed in order for him to reject H_0. By computation, he finds that his D must be .254 or larger for H_0 to be rejected, for

$$1.36 \sqrt{\frac{n_1 + n_2}{n_1 n_2}} = 1.36 \sqrt{\frac{55 + 60}{(55)(60)}} = .254$$

Large samples : one-tailed test. When n_1 and n_2 are large, and regardless of whether or not $n_1 = n_2$, we may make a one-tailed test by using

$$D = \text{maximum } [S_{n_1}(X) - S_{n_2}(X)] \qquad (6.10a)$$

We test the null hypothesis that the two samples have been drawn from the same population against the alternative hypothesis that the values of the population from which one of the samples was drawn are stochastically larger than the values of the population from which the other sample was drawn. For example, we may wish to test not simply whether an experimental group is different from a control group but whether the experimental group is "higher" than the control group.

It has been shown (Goodman, 1954) that

$$\chi^2 = 4D^2 \frac{n_1 n_2}{n_1 + n_2} \qquad (6.11)$$

has a sampling distribution which is approximated by the chi-square distribution with $df = 2$. That is, we may determine the significance of an observed value of D, as computed from formula (6.10a), by solving formula (6.11) for the observed values of D, n_1, and n_2, and referring to the chi-square distribution with $df = 2$ (Table C of the Appendix).

Example for Large Samples: One-tailed Test

In a study of correlates of authoritarian personality structure,[1] one hypothesis was that persons high in authoritarianism would show a greater tendency to possess stereotypes about members of various national and ethnic groups than would those low in authoritarianism. This hypothesis was tested with a group of 98 randomly selected college women. Each subject was given 20 photographs and asked to "identify" those whose nationality she recognized, by matching the appropriate photograph with the name of the national group. Subjects were free to "identify" (by matching) as many or as few photographs as they wished. Since, unknown to the subjects, all photographs were of Mexican nationals—either candidates for the Mexican legislature or winners in a Mexican beauty contest—and since the matching list of 20 different national and ethnic groups did not include "Mexican," the number of photographs which any subject "identified" constituted an index of that subject's tendency to stereotype.

Authoritarianism was measured by the well-known F scale of authoritarianism,[2] and the subjects were grouped as "high" and "low" scorers. "High" scorers were those who scored at or above the median on the F scale; "low" scorers were those who scored below the median. The prediction was that these two groups would differ in the number of photographs they "identified."

i. *Null Hypothesis.* H_0: women at this university who score low in authoritarianism stereotype as much ("identify" as many photographs) as women who score high in authoritarianism. H_1: women who score high in authoritarianism stereotype more ("identify" more photographs) than women who score low in authoritarianism.

ii. *Statistical Test.* Since the low scorers and the high scorers constitute two independent groups, a test for two independent samples was chosen. Because the number of photographs "identified" by a subject cannot be considered more than an ordinal measure of that subject's tendency to stereotype, a nonparametric test is desirable. The Kolmogorov-Smirnov two-sample test compares the two sample cumulative frequency distributions and determines whether the observed D indicates that they have been drawn from two populations, one of which is stochastically larger than the other.

---

[1] Siegel, S. 1954. Certain determinants and correlates of authoritarianism. *Genet. Psychol. Monogr.*, **49**, 187–229.

[2] Presented in Adorno, T. W., Frenkel-Brunswik, Else, Levinson, D. J., and Sanford, R. N. *The authoritarian personality.* New York: Harper, 1950.

iii. *Significance Level.* Let $\alpha = .01$. The sizes of n_1 and n_2 may be determined only after the data are collected, for subjects will be grouped according to whether they score at or above the median on the F scale or score below the median on the F scale.

iv. *Sampling Distribution.* The sampling distribution of

$$\chi^2 = \frac{4D^2(n_1 n_2)}{(n_1 + n_2)}$$

[i.e., formula (6.11)], where D is computed from formula (6.10a), is approximated by the chi-square distribution with $df = 2$. The probability associated with an observed value of D may be determined by computing χ^2 from formula (6.11) and referring to Table C.

v. *Rejection Region.* Since H_1 predicts the direction of the difference between the low and high F scorers, a one-tailed test is used. The region of rejection consists of all values of χ^2, as computed from formula (6.11), which are so large that the probability associated with their occurrence under H_0 for $df = 2$ is equal to or less than $\alpha = .01$.

vi. *Decision.* Of the 98 college women, 44 obtained F scores below the median. Thus $n_1 = 44$. The remaining 54 women obtained scores at or above the median: $n_2 = 54$. The number of photographs "identified" by each of the subjects in the two groups is given in Table 6.17. To apply the Kolmogorov-Smirnov test, we

TABLE 6.17. NUMBER OF LOW AND HIGH AUTHORITARIANS
"IDENTIFYING" VARIOUS NUMBERS OF PHOTOGRAPHS

Number of photographs "identified"	Low scorers	High scorers
0–2	11	1
3–5	7	3
6–8	8	6
9–11	3	12
12–14	5	12
15–17	5	14
18–20	5	6

recast these data into two cumulative frequency distributions, as in Table 6.18. For ease of computation, the fractions shown in Table 6.18 may be converted to decimal values; these values are shown in Table 6.19. By simple subtraction, we find the differences between the two sample distributions at the various intervals. The largest

TABLE 6.18. DATA IN TABLE 6.17 CAST FOR
KOLMOGOROV-SMIRNOV TEST

	Number of photographs "identified"						
	0–2	3–5	6–8	9–11	12–14	15–17	18–20
$S_{44}(X)$	$\frac{11}{44}$	$\frac{18}{44}$	$\frac{26}{44}$	$\frac{29}{44}$	$\frac{34}{44}$	$\frac{39}{44}$	$\frac{44}{44}$
$S_{54}(X)$	$\frac{1}{54}$	$\frac{4}{54}$	$\frac{10}{54}$	$\frac{22}{54}$	$\frac{34}{54}$	$\frac{38}{54}$	$\frac{54}{54}$

TABLE 6.19. DECIMAL EQUIVALENTS OF DATA IN TABLE 6.18

	Number of photographs "identified"						
	0–2	3–5	6–8	9–11	12–14	15–17	18–20
$S_{44}(X)$	.250	.409	.591	.659	.773	.886	1.0
$S_{54}(X)$	.018	.074	.185	.407	.630	.704	1.0
$S_{44}(X) - S_{54}(X)$	.232	.335	.406	.252	.143	.182	.0

of these differences in the predicted direction is .406. That is,

$$D = \text{maximum } [S_{n_1}(X) - S_{n_2}(X)] \qquad (6.10a)$$
$$= \text{maximum } [S_{44}(X) - S_{54}(X)]$$
$$= .406$$

With $D = .406$, we compute the value of χ^2 as defined by formula (6.11)

$$\chi^2 = 4D^2 \frac{n_1 n_2}{n_1 + n_2} \qquad (6.11)$$
$$= 4(.406)^2 \frac{(44)(54)}{44 + 54}$$
$$= 15.97$$

Reference to Table C reveals that the probability associated with $\chi^2 = 15.97$ for $df = 2$ is $p < .001$ (one-tailed test). Since this value is smaller than $\alpha = .01$, we may reject H_0 in favor of H_1.[*] We conclude that women who score high on the authoritarianism scale stereotype more ("identify" more photographs) than do women who score low on the scale.

It is interesting to notice that the chi-square approximation may also be used with small samples, but in this case it leads to a conservative

---

[*] Using a parametric test, Siegel made the same decision. He found that $t = 3.55$, $p < .001$ (one-tailed test).

test. That is, the error in the use of the chi-square approximation with small samples is always in the "safe" direction (Goodman, 1954, p. 168). In other words, if H_0 is rejected with the use of the chi-square approximation with small samples, we may surely have confidence in the decision. When this approximation is used for small samples, it is not necessary that n_1 and n_2 be equal.

To show how well the chi-square approximation works even for small samples, let us use it on the data presented in the example for small samples (above). In that case, $n_1 = n_2 = 10$, and D, as computed from formula (6.10a), was $\frac{7}{10}$. The chi-square approximation:

$$\chi^2 = 4D^2 \frac{n_1 n_2}{n_1 + n_2} \qquad (6.11)$$

$$= 4 \left(\frac{7}{10} \right)^2 \frac{(10)(10)}{10 + 10}$$

$$= 9.8$$

Table C shows that $\chi^2 = 9.8$ with $df = 2$ is significant at the .01 level. This is the same result as that which was obtained for these data by the use of Table L, which is based on exact computations.

Summary of procedure. These are the steps in the use of the Kolmogorov-Smirnov two-sample test:

1. Arrange each of the two groups of scores in a cumulative frequency distribution, using the same intervals (or classifications) for both distributions. Use as many intervals as are feasible.

2. By subtraction, determine the difference between the two sample cumulative distributions at each listed point.

3. By inspection, determine the largest of these differences; this is D. For a one-tailed test, D is the largest difference in the predicted direction.

4. The method for determining the significance of the observed D depends on the size of the samples and the nature of H_1:

   a. When $n_1 = n_2 = N$, and when $N \leq 40$, Table L is used. It gives critical values of K_D (the numerator of D) for various levels of significance, for both one-tailed and two-tailed tests.

   b. For a two-tailed test, when n_1 and n_2 are both larger than 40, Table M is used. In such cases it is not necessary that $n_1 = n_2$. Critical values of D for any given large values of n_1 and n_2 may be computed from the expressions given in the body of Table M.

   c. For a one-tailed test where n_1 and n_2 are large, the value of χ^2 with $df = 2$ which is associated with the observed D is computed from formula (6.11). The significance of the resulting value of χ^2 with $df = 2$ may be determined by reference to Table C. This chi-square approximation is also useful for small samples with $n_1 \neq n_2$, but in that application the test is conservative. If the observed value is equal to or larger than that given in the appropriate table for a particular level of significance, H_0 may be rejected at that level of significance.

Power-Efficiency

When compared with the t test, the Kolmogorov-Smirnov test has high power-efficiency (about 96 per cent) for small samples (Dixon, 1954). It would seem that as the sample size increases the power-efficiency would tend to decrease slightly.

The Kolmogorov-Smirnov test seems to be more powerful in all cases than either the χ^2 test or the median test.

The evidence seems to indicate that whereas for very small samples the Kolmogorov-Smirnov test is slightly more efficient than the Mann-Whitney test, for large samples the converse holds.

References

For other discussions of the Kolmogorov-Smirnov two-sample test, the reader may consult Birnbaum (1952; 1953), Dixon (1954), Goodman (1954), Kolmogorov (1941), Massey (1951a; 1951b), and Smirnov (1948).

THE WALD-WOLFOWITZ RUNS TEST

Function

The Wald-Wolfowitz runs test is applicable when we wish to test the null hypothesis that two independent samples have been drawn from the same population against the alternative hypothesis that the two groups differ in any respect whatsoever. That is, with sufficiently large samples the Wald-Wolfowitz test can reject H_0 if the two populations differ in any way: in central tendency, in variability, in skewness, or whatever. Thus it may be used to test a large class of alternative hypotheses. Whereas many other tests are addressed to particular sorts of differences between two groups (e.g., the median test determines whether the two samples have been drawn from populations with the same median), the Wald-Wolfowitz test is addressed to any sort of difference.

Rationale and Method

The Wald-Wolfowitz test assumes that the variable under consideration has an underlying distribution which is continuous. It requires that the measurement of that variable be in at least an ordinal scale.

To apply the test to data from two independent samples of size n_1 and n_2, we rank the $n_1 + n_2$ scores in order of increasing size. That is, we cast the scores of all subjects in both groups into one ordering.   Then we determine the number of runs in this ordered series.   A run is defined as any sequence of scores from the same group (either group 1 or group 2).

For example, suppose we observed these scores from group A (consisting of 3 cases—$n_1 = 3$) and group B (consisting of 4 cases—$n_2 = 4$):

Scores for group A	12	16	8	
Scores for group B	11	6	6	3

When these 7 scores are cast in one ordered series, we have:

3	6	6	8	11	12	16
B	B	B	A	B	A	A

Notice that we retain the identity of each score by accompanying that score with the sign of the group to which it belongs.   We then observe the order of the occurrence of these signs (A's and B's) to determine the number of runs.   Four runs occurred in this series: the 3 lowest scores were all from group B and thus constituted 1 run of B's; the next highest score is a run of a single A; another run constituted by 1 B follows; and the two highest scores are both from group A and constitute the final run.

Now we may reason that if the two samples are from the same population (that is, if H_0 is true), then the scores of the A's and the B's will be well mixed.   In that case r, the number of runs, will be relatively large.   It is when H_0 is false that r is small.

For example, r will be small if the two samples were drawn from populations having different medians.   Suppose the population from which the A cases were drawn had a higher median than the population from which the B cases were drawn.   In the ordered series of scores from the two samples, we would expect a long run of B's at the lower end of the series and a long run of A's at the upper end, and consequently an r which is relatively small.

Again, suppose the samples were drawn from populations which differed in variability.   If the population from which the A cases were drawn was highly dispersed, whereas the population from which the B cases were drawn was homogeneous or compact, we would expect a long run of A's at each end of the ordered series and thus a relatively small value of r.

Similar arguments can be presented to show that when the populations from which the n_1 and n_2 cases were drawn differ in skewness or kurtosis, then the size of r will also be "too small," i.e., small relative to the sizes of n_1 and n_2.

In general, then, we reject H_0 if r = the number of runs is "too small." The sampling distribution of r arises from the fact that when two different kinds of objects (say n_1 and n_2) are arranged in a single line, the total number of different possible arrangements is

$$\binom{n_1 + n_2}{n_1} = \binom{n_1 + n_2}{n_2}$$

From this it can be shown (Stevens, 1939; Mood, 1950, pp. 392–393) that the probability of getting an observed value of r or an even smaller value is

$$p(r \leq r') = \frac{1}{\binom{n_1 + n_2}{n_1}} \sum_{r=2}^{r'} (2) \binom{n_1 - 1}{\frac{r}{2} - 1} \binom{n_2 - 1}{\frac{r}{2} - 1} \quad (6.12a)$$

when r is an even number. When r is an odd number, that probability is given by

$$p(r \leq r') = \frac{1}{\binom{n_1 + n_2}{n_1}} \sum_{r=2}^{r'} \left[\binom{n_1 - 1}{k - 1} \binom{n_2 - 1}{k - 2} \right.$$
$$\left. + \binom{n_1 - 1}{k - 2} \binom{n_2 - 1}{k - 1} \right] \quad (6.12b)$$

where $r = 2k - 1$.

Small samples. Tables of critical values of r, based on formulas (6.12a) and (6.12b), have been constructed. Table F_I of the Appendix presents critical values of r for n_1, $n_2 \leq 20$. These values are significant at the .05 level. That is, if an observed value of r is equal to or less than the value tabled for the observed values of n_1 and n_2, H_0 may be rejected at the .05 level of significance. If the observed value of r is larger than that shown in Table F_I, we can only conclude that *in terms of the total number of runs observed*, the null hypothesis cannot be rejected at $\alpha = .05$.

Example for Small Samples

Twelve four-year-old boys and twelve four-year-old girls were observed during two 15-minute play sessions, and each child's play during both periods was scored for incidence of and degree of agression.[1] With these scores, it is possible to test the hypothesis that there are sex differences in the amount of aggression shown.

    i. *Null Hypothesis.* H_0: incidence and degree of aggression are the same in four-year-olds of both sexes. H_1: four-year-old boys

[1] Siegel, Alberta E. 1956. Film-mediated fantasy aggression and strength of aggressive drive. *Child Develpm.*, **27**, 365–378.

and four-year-old girls display differences in incidence and degree of aggression.

ii. *Statistical Test.* Since the data are in an ordinal scale, and since the hypothesis concerns differences *of any kind* between the aggression scores of two independent groups (boys and girls), the Wald-Wolfowitz runs test is chosen.

iii. *Significance Level.* Let $\alpha = .05$. $n_1 = 12 = $ the number of boys, and $n_2 = 12 = $ the number of girls.

iv. *Sampling Distribution.* From the sampling distribution of r, critical values have been tabled in Table F_I for n_1, $n_2 \leq 20$. (Although $n_1 = n_2$ in this example, this is not necessary for the use of the runs test.)

v. *Rejection Region.* The region of rejection consists of all values of r which (for $n_1 = 12$ and $n_2 = 12$) are so small that the probability associated with their occurrence under H_0 is equal to or less than $\alpha = .05$.

vi. *Decision.* Each child's score for his total aggression in both sessions was obtained. These scores are given in Table 6.20.

TABLE 6.20. AGGRESSION SCORES OF BOYS AND GIRLS IN FREE PLAY

Boys	Girls
86	55
69	40
72	22
65	58
113	16
65	7
118	9
45	16
141	26
104	36
41	20
50	15

Now if we combine the scores of the boys (B's) and girls (G's) in a single ordered series, we may determine the number of runs of G's and B's. This ordered series is shown in Table 6.21. Each run is underlined, and we observe that $r = 4$.

Reference to Table F_I reveals that for $n_1 = 12$ and $n_2 = 12$, an r of 7 is significant at the .05 level. Since our value of r is smaller than that tabled, we may reject H_0 at $\alpha = .05$.* We conclude that boys and girls display differences in aggression in the free play situation.

---

* Using the nonparametric Mann-Whitney U test for the data shown in Table 6.20, the investigator rejected H_0 at the $p < .0002$ level (two-tailed test).

TABLE 6.21. DATA IN TABLE 6.20 CAST FOR RUNS TEST

Score	7	9	15	16	16	20	22	26	36	40
Group	G	G	G	G	G	G	G	G	G	G
Run					1					

Score	41	45	50	55	58
Group	B	B	B	G	G
Run		2		3	

Score	65	65	69	72	86	104	113	118	141
Group	B	B	B	B	B	B	B	B	B
Run					4				

Large samples. When either n_1 or n_2 is larger than 20, Table F_I cannot be used. However, for such large samples the sampling distribution under H_0 for r is approximately normal, with

$$\text{Mean} = \mu_r = \frac{2n_1 n_2}{n_1 + n_2} + 1$$

and   Standard deviation $= \sigma_r = \sqrt{\frac{2n_1 n_2 (2n_1 n_2 - n_1 - n_2)}{(n_1 + n_2)^2 (n_1 + n_2 - 1)}}$

That is, the expression

$$z = \frac{r - \mu_r}{\sigma_r} = \frac{r - \left(\dfrac{2n_1 n_2}{n_1 + n_2} + 1\right)}{\sqrt{\dfrac{2n_1 n_2 (2n_1 n_2 - n_1 - n_2)}{(n_1 + n_2)^2 (n_1 + n_2 - 1)}}} \tag{6.13}$$

is approximately normally distributed with zero mean and unit variance. Thus Table A of the Appendix, which gives the probability associated with the occurrence under H_0 of values as extreme as an observed z, may be used with large samples to determine the significance of an observed value of r.

A correction for continuity should be used when $n_1 + n_2$ is not very large. The correction is required because the distribution of empirical values of r must of necessity be discrete, whereas with large samples we approximate that sampling distribution by the normal curve, a continuous curve. This approximation can be improved by correcting for continuity. The correction is achieved by subtracting .5 from the absolute difference between r and μ_r:

$$z = \frac{|r - \mu_r| - .5}{\sigma_r} \tag{6.14}$$

Thus to compute the value of z with the correction for continuity incorporated, we use formula (6.14):

$$z = \frac{\left| r - \left(\frac{2n_1 n_2}{n_1 + n_2} + 1 \right) \right| - .5}{\sqrt{\frac{2n_1 n_2 (2n_1 n_2 - n_1 - n_2)}{(n_1 + n_2)^2 (n_1 + n_2 - 1)}}} \qquad (6.14)$$

Computation of formula (6.14) will yield a z whose associated tabular value (in Table A) gives the probability under H_0 of a value as small as the observed value of r.   If the z obtained from the use of formula (6.14) has an associated probability, p (read directly from Table A), which is equal to or less than α, then H_0 may be rejected at the α level of significance.

Example for Large Samples

In a study which tested the equipotentiality theory,[1] Ghiselli compared the learning (in a brightness-discrimination task) of 21 normal rats with the relearning of 8 postoperative rats with cortical lesions.   That is, the number of trials to relearning required postoperatively by the 8 E rats was compared with the number of trials to learning required by the 21 C rats.

i. *Null Hypothesis.* H_0: there is no difference between normal rats and postoperative rats with cortical lesions with respect to rate of learning (or relearning) in the brightness-discrimination task. H_1: the two groups of rats differ with respect to rate of learning (or relearning).

ii. *Statistical Test.*   The Wald-Wolfowitz test was chosen to provide an over-all test for differences between the two groups. Since $n_2 > 20$, the normal curve approximation will be used.   And since $n_1 + n_2 = 29$ is fairly small, the correction for continuity will be employed, i.e., formula (6.14) will be used.

iii. *Significance Level.*   Let $\alpha = .01$. $n_1 = 8$ postoperative rats and $n_2 = 21$ normal rats.

iv. *Sampling Distribution.*   Table A gives the probability associated with the occurrence under H_0 of a value as extreme as any z computed from formula (6.14).

v. *Rejection Region.*   The region of rejection consists of all values of z which are so extreme that the probability associated with their occurrence under H_0 is equal to or less than $\alpha = .01$.

vi. *Decision.*   Table 6.22 gives the number of trials to relearning required by the 8 postoperative animals and the number of trials to

---

[1] Ghiselli, E. E.   1938.   Mass action and equipotentiality of the cerebral cortex in brightness discrimination.   *J. Comp. Psychol.*, **25**, 273–290.

TABLE 6.22. TRIALS TO LEARNING (RELEARNING) REQUIRED BY E
AND C RATS

E Rats	C Rats
20	23
55	8
29	24
24	15
75	8
56	6
31	15
45	15
	21
	23
	16
	15
	24
	15
	21
	15
	18
	14
	22
	15
	14

learning required by the 21 normal animals. From these scores we
may determine the number of runs; the runs are shown in Table
6.23. We see that $r = 6$.

TABLE 6.23. DATA IN TABLE 6.22 CAST FOR RUNS TEST

Score	6	8	8	14	14	15	15	15	15	15	15	15	16	18
Group	C	C	C	C	C	C	C	C	C	C	C	C	C	C
Run							1							

Score	20	21	21	22	23	23	24	24	24
Group	E	C	C	C	C	C	C	E	C
Run	2			3				4	5

Score	29	31	45	55	56	75
Group	E	E	E	E	E	E
Run			6			

To determine the probability under H_0 of such a small value or even
smaller value of r, we compute the value of z, substituting our
observed values ($r = 6$, $n_1 = 8$, and $n_2 = 21$) in formula (6.14):

$$z = \frac{\left| r - \left(\dfrac{2n_1 n_2}{n_1 + n_2} + 1 \right) \right| - .5}{\sqrt{\dfrac{2n_1 n_2 (2n_1 n_2 - n_1 - n_2)}{(n_1 + n_2)^2 (n_1 + n_2 - 1)}}} \tag{6.14}$$

$$= \frac{\left| 6 - \left(\dfrac{(2)(8)(21)}{8 + 21} + 1 \right) \right| - .5}{\sqrt{\dfrac{(2)(8)(21)[(2)(8)(21) - 8 - 21]}{(8 + 21)^2 (8 + 21 - 1)}}}$$

$$= 2.92$$

Reference to Table A indicates that $z \geq 2.92$ has probability of occurrence under H_0 of $p = .0018$. Since this value of p is smaller than $\alpha = .01$, our decision is to reject H_0 in favor of H_1.* We conclude that the two groups of animals differ significantly in their rate of learning (relearning).

Ties. Ideally no ties should occur in the scores used for a runs test, inasmuch as the populations from which the samples were drawn are assumed to be continuous distributions. In practice, however, inaccurate or insensitive measurement results in the occasional occurrence of ties. When ties occur between members of the different groups, then the sequence of scores is not unique. That is, suppose three subjects obtain tied scores. Two of these are A's and one is a B. In making the ordered series of scores, how should we group these three? If we group them as $A\ B\ A$, then we will have a different number of runs than if we group them as $A\ A\ B$ or (alternatively) as $B\ A\ A$.

If all ties are within the same sample, then the number of runs (r) is unaffected and therefore the obtained significance level is unaffected. But if observations from one sample are tied with observations from the other sample, we cannot obtain a unique ordered series and therefore usually cannot obtain a unique value of r, as we have just shown.

This problem occurred in the example just presented. Three rats required 24 trials to learn to the criterion. In Table 6.23 we ordered these cases as $C\ E\ C$. We might just as well have ranked them $E\ C\ C$. As it happens, no matter what order we had used, in this case, r would have been 6 or smaller, and thus our decision would have been to reject H_0 in any case. For this reason ties presented no major problem in reaching a statistical decision concerning those data.

In other sets of data, they might. Our procedure with ties is to

---

* Using a parametric test, Ghiselli reached the same decision. He reported a critical ratio of 3.95, which would allow him to reject H_0 at $\alpha = .00005$.

break the ties in all possible ways and observe the resulting values of *r*. If all these values are significant with respect to the previously set value of α, then ties present no major problem, although they do increase the tedium of computation.

If the various possible ways of breaking up ties lead to some values of *r* which are significant and some which are not, the decision is more difficult. In this case, we suggest that the researcher determine the probability of occurrence associated with each possible value of *r* and take the average of these *p*'s as his obtained probability for use in deciding whether to accept or reject H_0.

If the number of ties between scores in the two different samples is large, *r* is essentially indeterminate. In such cases, the Wald-Wolfowitz test is inapplicable.

Summary of procedure. These are the steps in the use of the Wald-Wolfowitz runs test:

1. Arrange the $n_1 + n_2$ scores in a single ordered series.
2. Determine *r* = the number of runs.
3. The method for determining the significance of the observed value of *r* depends on the size of n_1 and n_2:

   a. If both n_1 and n_2 are 20 or smaller, Table F_1 gives critical values of *r* at the .05 level of significance. If the observed value of *r* is equal to or smaller than that tabled for the observed values of n_1 and n_2, then H_0 may be rejected at $\alpha = .05$.

   b. If either n_1 or n_2 is larger than 20, formula (6.13) or (6.14) may be used to compute the value of *z* whose associated probability under H_0 may be determined by reading the *p* associated with that *z*, as given in Table A. Choose formula (6.14) if $n_1 + n_2$ is not very large and thus a correction for continuity is desirable. If the *p* is equal to or less than α, reject H_0.

4. If ties occur between scores from the two different samples, follow the procedure suggested above in the discussion of ties.

Power-Efficiency

Little is known about the power-efficiency of the Wald-Wolfowitz test. Moses (1952a) points out that statistical tests which test H_0 against many alternatives simultaneously—and the runs test is such a test—are not very good at guarding against accepting H_0 erroneously with respect to any one particular alternative.

For instance, if we were interested simply in testing whether two samples come from populations with the same location, the Mann-Whitney *U* test would be a more powerful test than the runs test because it is specifically designed to disclose differences of this type, whereas the runs test is designed to disclose differences of any type and is thus less powerful in disclosing any particular kind. This difference was illustrated in the example for small samples shown above. The investigator was interested in sex differences in location of aggression scores, and therefore used the U test. We tested the data for differences of any sort, using the runs test. Both tests rejected H_0, but the Mann-Whitney U test did so at a much more extreme level of significance.

Mood (1954) points out that when the Wald-Wolfowitz test is used to test H_0 against specific alternatives regarding location or variability, it has theoretic asymptotic efficiency of zero. However, Lehmann (1953) discusses whether it is proper to apply the notion of asymptotic normality to the runs test.

Smith (1953) states that empirical evidence indicates that the power-efficiency of the Wald-Wolfowitz test is about 75 per cent for sample sizes near 20.

References

The reader may find discussions of the runs test in Lehmann (1953), Moses (1952a), Smith (1953), Stevens (1939), and Swed and Eisenhart (1943).

THE MOSES TEST OF EXTREME REACTIONS

Function and Rationale

In the behavioral sciences, we sometimes expect that an experimental condition will cause some subjects to show extreme behavior in one direction while it causes others to show extreme behavior in the opposite direction. Thus we may think that economic depression and political instability will cause some people to become extremely reactionary and others to become extremely "left-wing" in their political opinions. Or we may expect environmental unrest to create extreme excitement in some mentally ill people while it creates extreme withdrawal in others. In psychological research utilizing the perception-centered approach to personality, there are theoretical reasons to predict that "perceptual defense" may manifest itself in either an extremely rapid "vigilant" perceptual response or an extremely slow "repressive" perceptual response.

The Moses test is specifically designed for use with data (measured in at least an ordinal scale) collected to test such hypotheses. It should be used when it is expected that the experimental condition will affect some subjects in one way and others in the opposite way. In studies of perceptual defense, for example, we expect the control subjects to evince "medium" or "normal" responses, while we expect the experimental subjects to give either "vigilant" or "repressive" responses, thus getting either high or low scores in comparison to those of the controls.

In such studies, statistical tests addressed to differences in central tendency will shield rather than reveal group differences. They lead to acceptance of the null hypothesis when it should be rejected, because when some of the experimental subjects show "vigilant" responses and thus obtain very low latency scores while others show "repressive" responses and thus obtain very high latency scores, the average of the scores of the experimental group may be quite close to the average score of controls (all of whom may have obtained scores which are "medium").

Although the Moses test is specifically designed for the sort of data mentioned above, it is also applicable when the experimenter expects that one group will score low and the other group will score high. However, Moses (1952b) points out that in such cases a test based on medians or on mean ranks, e.g., the Mann-Whitney U test, is more efficient and is therefore to be preferred to the Moses test. The latter test is uniquely valuable when there exist a priori grounds for believing that the experimental condition will lead to extreme scores in either direction.

The Moses test focuses on the *span* or spread of the control cases. That is, if there are n_C control cases and n_E experimental cases, and the $n_E + n_C$ scores are arranged in order of increasing size, and if the null hypothesis (that the E's and C's come from the same population) is true, then we should expect that the E's and C's will be well mixed in the ordered series. We should expect under H_0 that some of the extremely high scores will be E's and some C's, that some of the extremely low scores will be E's and some C's, and that the middle range of scores would include a mixture of E's and C's. However, if the alternative hypothesis (that the E scores represent defensive responses) is true, then we would expect that (a) most of the E scores will be low, i.e., "vigilant," or (b) most of the E scores will be high, i.e., "repressive," or (c) a considerable proportion of the E's will score low and another considerable proportion will score high, i.e., some E responses will be "vigilant" while others are "repressive." In any of these three cases, the scores of the C's will be unduly congested and consequently their span will be relatively small. If situation (a) holds, then the C's will be congested at the high end of the series, if (b) holds the C's will be congested at the low end of the series, and if (c) holds the C's will be congested in the middle of the ordered series. The Moses test determines whether the C scores are so closely compacted or congested relative to the $n_E + n_C$ scores as to call for rejecting the null hypothesis that both E's and C's come from the same population.

Method

To compute the Moses test, combine the scores from the E and C groups, and arrange these scores in a single ordered series, retaining the group identity of each score.

Then determine the span of the C scores by noting the lowest and the highest C scores and counting the number of cases between them, including both extremes. That is, the span, symbolized as s', is defined as the smallest number of consecutive scores in an ordered series necessary to include all the C scores. For ease of computation, we may rank each score and determine s' from the ordered series of the ranks assigned to the $n_E + n_C$ cases.

For example, suppose scores are obtained for $n_C = 6$ and $n_E = 7$ cases. When these 13 cases are ranked together, we have this series:

Rank	1	2	3	4	5	6	7	8	9	10	11	12	13
Group	E	E	C	E	C	E	C	C	C	E	C	E	E

The span of the C scores in this case extends over 9 ranks (from 3 to 11 inclusive) and thus $s' = 9$.

Notice that in general s' is equal to the difference between the extreme C ranks plus 1. In the present case, $s' = 11 - 3 + 1 = 9$.

The Moses test determines whether the observed value of s' is too small a value to be thought to have reasonably arisen by chance if the E's and C's are from the same population. That is, the sampling distribution of s' under the null hypothesis is known (Moses, 1952b) and may be used for tests of significance.

The reader will have observed that s' is essentially the *range* of the C scores, and he may object that the well-known instability of the range makes s' an unreliable index to the actual spread or compactness of the C scores. Moses points out that it is usually necessary to modify s' in order to take care of just this problem. The modification is especially important when n_C is large, because especially in this case is the range (span) of C's an inefficient index to the spread of the group, due to possible sampling fluctuations.

The modification suggested by Moses is that the researcher, in advance of collecting his data, arbitrarily select some small number, h. After the data are collected, he may subtract h control scores from both extremes of the range of control scores. The span is found for those scores which remain. That is, the span is found after h control scores have been dropped from each extreme of the series.

For example, in the data given earlier, the experimenter might have decided in advance that $h = 1$. Then he would have dropped ranks 3 and 11 from the C scores before determining the span. In that case, the "truncated span," symbolized as s_h, would be $s_h = 9 - 5 + 1 = 5$. This is given as: $s_h = 5$, $h = 1$. Thus s_h is defined as the smallest number of consecutive ranks necessary to include all the control scores except the h least and the h greatest of them.

Notice that s_h can never be smaller than $n_C - 2h$ and can never be larger than $n_C + n_E - 2h$. The sampling distribution, then, should tell us the probability under H_0 of observing an s_h which exceeds the minimum value $(n_C - 2h)$ by any specified amount.

If we use g to represent the amount by which an observed value of s_h exceeds $n_C - 2h$, we may determine the probability under H_0 of observing a particular value of s_h or less as

$$p(s_h \leq n_C - 2h + g) = \frac{\sum_{i=0}^{g} \binom{i + n_C - 2h - 2}{i} \binom{n_E + 2h + 1 - i}{n_E - i}}{\binom{n_C + n_E}{n_C}}$$

(6.15)

Thus for any observed values of n_C and n_E and a given previously set value of h, one first finds the minimum possible truncated span: $n_C - 2h$. Then one finds the value of $g =$ the amount that the observed s_h exceeds the value of $(n_C - 2h)$. The probability of the occurrence of the observed value of s_h or less under H_0 is found by cumulating the terms in the numerator of formula (6.15). If $g = 1$, then one must sum the numerator terms for $i = 0$ and $i = 1$. If $g = 2$, then one must sum three numerator terms: for $i = 0$, $i = 1$, and $i = 2$. The computations called for by formula (6.15) are illustrated in the following example of the use of the Moses test.

Example

In a pilot study of the perception of interpersonal hostility in film dramas, the experimenter[1] compared the amount of hostility perceived by two groups of female subjects. The E group were women whose personality test data revealed that they had difficulty in handling their own aggressive impulses. The C group were women whose personality tests revealed that they had little or no disturbance in the area of aggression and hostility. Each of the 9 E subjects and the 9 C subjects was shown a filmed drama and asked to rate the amount of aggression and hostility shown by the characters in the drama.

[1] This example cites unpublished pilot study data made available to the author through the courtesy of the experimenter, Dr. Ellen Tessman.

The hypothesis was that the E subjects would *either* underattribute *or* overattribute hostility to the film characters. Underattribution is indicated by a low score, whereas overattribution is indicated by a high score. It was predicted that the C subjects' scores would be more moderate than those of the E subjects, i.e., that the C's would evince less distortion in their perception of interpersonal hostility.

i. *Null Hypothesis.* H_0: women who have personal difficulty in handling aggressive impulses do not differ from women with relatively little disturbance in this area in the amount of hostility that they attribute to the film characters. H_1: women who have personal difficulty in handling aggressive impulses are *more extreme* than others in their judgments of hostility in film characters—some underattribute and others overattribute.

ii. *Statistical Test.* Since defensive (extreme) reactions are being predicted, and since the study employs two independent groups, the Moses test is appropriate for an analysis of the research data. In advance of collecting the data, the researcher set h at 1.

iii. *Significance Level.* Let $\alpha = .05$. $n_E = 9$ and $n_C = 9$.

iv. *Sampling Distribution.* The probability associated with the occurrence under H_0 of any value as small as an observed s_h is given by formula (6.15).

v. *Rejection Region.* The region of rejection consists of all values of s_h which are so small that the probability associated with their occurrence under H_0 is equal to or less than $\alpha = .05$.

vi. *Decision.* The scores for attribution of aggression by the E and C subjects are given in Table 6.24, which also shows the rank of

TABLE 6.24. ATTRIBUTION OF AGGRESSION TO CHARACTERS IN FILM

E Subjects		C Subjects	
Score	Rank	Score	Rank
25	18	12	10
5	3	16	15
14	13	6	4
19	17	13*	12
0	1	13*	11
17	16	3	2
15	14	10*	7
8*	6	10*	8
8*	5	11	9

* When ties occur between two members of the same group, the value of s_h is unaffected and thus the use of tied ranks is unnecessary. For a discussion of the problem of ties in the Moses test, see the section following this example.

each. When these ranks are ordered in a single series, we have the data shown in Table 6.25.

TABLE 6.25. DATA IN TABLE 6.24 CAST FOR MOSES TEST

Rank	1	2	3	4	5	6	7	8	9	10	11	12	13	14	15	16	17	18
Group	E	C	E	C	E	E	C	C	C	C	C	C	E	E	C	E	E	E

Since $h = 1$, the most extreme rank at each end of the C range is dropped; these are ranks 2 and 15. Without these two ranks, the truncated span of the C scores is 9. That is,

$$s_h = 9 \qquad h = 1$$

Now the minimum possible s_h would be $(n_C - 2h) = 9 - 2 = 7$. Thus the amount by which the observed s_h exceeds the minimum possible is $9 - 7 = 2$. Thus $g = 2$. To determine the probability of occurrence under H_0 of $s_h \le 9$ when $n_C = 9$, $n_E = 9$, and $g = 2$, we substitute these values into formula (6.15):

$$
\begin{aligned}
p(s_h \le n_C - 2h + g) &= \frac{\displaystyle\sum_{i=0}^{g} \binom{i + n_C - 2h - 2}{i}\binom{n_E + 2h + 1 - i}{n_E - i}}{\dbinom{n_C + n_E}{n_C}} \\[2em]
&= \frac{\displaystyle\sum_{i=0}^{2} \binom{i + 9 - 2 - 2}{i}\binom{9 + 2 + 1 - i}{9 - i}^{*}}{\dbinom{9 + 9}{9}} \\[2em]
&= \frac{\dbinom{5}{0}\dbinom{12}{9} + \dbinom{6}{1}\dbinom{11}{8} + \dbinom{7}{2}\dbinom{10}{7}}{\dbinom{18}{9}} \\[2em]
&= \frac{(1)(220) + (6)(165) + (21)(120)}{48,620} \\[1em]
&= .077
\end{aligned}
$$

* For any positive integers, say a and b,

$$\binom{a}{b} = \frac{a!}{b!\,(a - b)!} \qquad \text{if } a \ge b$$

and

$$\binom{a}{b} = 0 \qquad \text{if } a < b$$

Table T of the Appendix gives numerical values for binomial coefficients $\dbinom{N}{x}$ for $N \le 20$.

Since $p = .077$ is larger than $\alpha = .05$, the data do not permit us to reject H_0 at our previously set level of significance. We conclude that, on the basis of these data, we cannot say at $\alpha = .05$ that the E subjects differ significantly from the C subjects in their attribution of aggression to the film characters. The p is sufficiently small, however, to be considered "promising" in pilot study data such as these.

Ties. Tied observations between two or more members of the same group do not affect the value of s_h. When there are tied observations between members of the two different groups, however, there may be more than one value of s_h, depending on how the tie is broken. When this is the case, the researcher should break the ties in his data in *all* possible ways, find the probability under H_0 associated with *each* such break [by using formula (6.15) for each], and take the average of these probabilities as the one to use in making his decision about H_0. If the number of ties between groups is large, then the Moses test is inapplicable.

Summary of procedure. These are the steps in the use of the Moses test:

1. In advance of the collection of data, specify the value of h.

2. When the scores have been collected, rank them in a single series, retaining the group identity of each rank.

3. Determine the value of s_h, the span of the control ranks after the h most extreme C ranks at each end of the series have been dropped.

4. Determine the value of g, the amount by which the observed value of s_h exceeds $n_C - 2h$.

5. Determine the probability associated with the observed data by computing the value of p as given by formula (6.15). If ties occurred between groups, break them in all possible ways and find the p for each such break; the average of these p's is used as the p in the decision.

6. If p is equal to or smaller than α, reject H_0.

Power

The power of the Moses test has not been reported. However, when the test is used for its special purpose (i.e., for testing the hypothesis that the members of one group will be extreme with respect to the members of another group), it is more efficient than tests that are sensitive only to shifts in location (central tendency) or in dispersion. Of course, as we have pointed out earlier, if the hypothesis under test deals specifically with central tendencies, then a test based on medians or mean ranks, e.g., the Mann-Whitney U test, will make more efficient use of the information in the data.

References

Further information on this test is contained in Moses (1952b).

THE RANDOMIZATION TEST FOR TWO INDEPENDENT SAMPLES

Function

The randomization test for two independent samples is a useful and powerful nonparametric technique for testing the significance of the difference between the means of two independent samples when n_1 and n_2 are small. The test employs the numerical values of the scores, and therefore requires at least interval measurement of the variable being studied. With the randomization test we can determine the exact probability under H_0 associated with our observations, and can do so without assuming normal distributions or homogeneity of variance in the populations involved (which must be assumed if the parametric equivalent, the t test, is used).

Rationale and Method

Consider the case of two small independent samples, either drawn at random from two populations or arising from the random assignment of two treatments to the members of a group whose origins are arbitrary. Group A includes 4 subjects; $n_1 = 4$. Group B includes 5 subjects; $n_2 = 5$. We observe the following scores:

Scores for group A	0	11	12	20	
Scores for group B	16	19	22	24	29

With these scores,[1] we wish to test the null hypothesis of no difference between the means against the alternative hypothesis that the mean of the population from which group A was drawn is smaller than the mean of the population from which group B was drawn.

Now under the null hypothesis, all n_1 and n_2 observations are from the same population. That is, it is merely a matter of chance that certain scores are labeled A and others are labeled B. The assignment of the labels A and B to the scores in the particular way observed may be conceived as one of many equally likely accidents if H_0 is true. Under H_0, the labels could have been assigned to the scores in any of 126 equally

[1] This example is taken from Pitman, E. J. G. 1937a. Significance tests which may be applied to samples from any populations. Supplement to *J. Royal Statist. Soc.*, **4**, 122.

likely ways:

$$\binom{n_1 + n_2}{n_1} = \binom{4 + 5}{4} = 126$$

Under H_0, only once in 126 trials would it happen that the four smallest scores of the nine would all acquire the label A, while the five largest acquired the label B.

Now if just such a result should occur in an actual single-trial experiment, we could reject H_0 at the $p = \frac{1}{126} = .008$ level of significance, applying the reasoning that if the two groups were really from a common population, i.e., if H_0 were really true, there is no good reason to think that the most extreme of 126 possible outcomes should occur on just the trial that constitutes our experiment. That is, we would decide that there is little likelihood that the observed event could occur under H_0, and therefore we would reject H_0 when the event did occur. This is part of the familiar logic of statistical inference.

The randomization test specifies a number of the most extreme possible outcomes which could occur with $n_1 + n_2$ scores, and designates these as the region of rejection. When we have $\binom{n_1 + n_2}{n_1}$ equally likely occurrences under H_0, for some of these the difference between ΣA (the sum of group A's scores) and ΣB (the sum of group B's scores) will be extreme. The cases for which these differences are largest constitute the region of rejection.

If α is the significance level, then the region of rejection consists of the $\alpha \binom{n_1 + n_2}{n_1}$ most extreme of the possible occurrences. That is, the *number* of possible outcomes constituting the region of rejection is $\alpha \binom{n_1 + n_2}{n_1}$. The particular outcomes chosen to constitute that number are those outcomes for which the difference between the mean of the A's and the mean of the B's is largest. These are the occurrences in which the difference between ΣA and ΣB is greatest. Now if the sample we obtain is among those cases listed in the region of rejection, we reject H_0 at significance level α.

In the example of 9 scores given above, there are $\binom{n_1 + n_2}{n_1} = 126$ possible differences between ΣA and ΣB. If $\alpha = .05$, then the region of rejection consists of $\alpha \binom{n_1 + n_2}{n_1} = .05(126) = 6.3$ extreme outcomes. Since the alternative hypothesis is directional, the region of rejection consists of the 6 most extreme possible outcomes in the specified direction.

Under the alternative hypothesis that $\mu_A < \mu_B$, the 6 most extreme possible outcomes constituting the region of rejection of $\alpha = .05$ (one-tailed test) are those given in Table 6.26. The third of these possible

TABLE 6.26. THE SIX MOST EXTREME POSSIBLE OUTCOMES IN THE
PREDICTED DIRECTION
(These constitute the region of rejection for the randomization test when $\alpha = .05$)

Possible scores for 5 B cases	Possible scores for 4 A cases	$\Sigma B - \Sigma A$
19 20 22 24 29	0 11 12 16	$114 - 39 = 75$
16 20 22 24 29	0 11 12 19	$111 - 42 = 69$
16 19 22 24 29	0 11 12 20	$110 - 43 = 67^*$
16 19 20 24 29	0 11 12 22	$108 - 45 = 63$
12 20 22 24 29	0 11 16 19	$107 - 46 = 61$
16 19 20 22 29	0 11 12 24	$106 - 47 = 59$

* The sample obtained.

extreme outcomes, the one with an asterisk, is the sample we obtained. Since our observed scores are in the region of rejection, we may reject H_0 at $\alpha = .05$. The exact probability (one-tailed) of the occurrence of the observed scores or a set more extreme under H_0 is $p = \frac{3}{126} = .024$.

Now if the alternative hypothesis had not predicted the direction of the difference, then of course a two-tailed test of H_0 would have been in order. In that case, the 6 sets of possible outcomes in the region of rejection would consist of the 3 most extreme possible outcomes in one direction and the 3 most extreme possible outcomes in the other direction. It would include the 6 possible outcomes whose difference between ΣA and ΣB was greatest in absolute value. For illustrative purposes, the 6 most extreme possible outcomes for a two-tailed test at $\alpha = .05$ for the 9 scores presented earlier are given in Table 6.27. With our observed scores H_0 would have been rejected in favor of the alternative hypothesis that $\mu_A \neq \mu_B$, because the obtained sample (shown with asterisk in Table 6.27) is one of the 6 most extreme of the possible outcomes in either direction. The exact probability (two-tailed) associated with the occurrence under H_0 of a set as extreme as the one observed is $p = \frac{6}{126} = .048$.

Large samples. When n_1 and n_2 are large, the computations necessary for the randomization test may be extremely tedious. However, they may be avoided, for Pitman has shown (1937a) that for n_1 and n_2 large, if the kurtosis of the combined samples is small and if the ratio of n_1 to n_2 lies between $\frac{1}{5}$ and 5, that is, if the larger sample is not more than five

TABLE 6.27. THE SIX MOST EXTREME POSSIBLE OUTCOMES IN
EITHER DIRECTION
(These constitute the two-tailed region of rejection for the randomization test
when $\alpha = .05$)

Possible scores for 5 B cases					Possible scores for 4 A cases				$|\Sigma B - \Sigma A|$
19	20	22	24	29	0	11	12	16	$|114 - 39| = 75$
0	11	12	16	19	20	22	24	29	$|58 - 95| = 37$
16	20	22	24	29	0	11	12	19	$|111 - 42| = 69$
0	11	12	16	20	19	22	24	29	$|59 - 94| = 35$
16	19	22	24	29	0	11	12	20	$|110 - 43| = 67^*$
0	11	12	16	22	19	20	24	29	$|61 - 92| = 31$

* The sample obtained.

times larger than the smaller sample, then the randomization distribution of $\left(\dfrac{n_1 + n_2}{n_2}\right)$ possible outcomes is closely approximated by the t distribution. That is, if the above-mentioned two conditions (small kurtosis and $\dfrac{1}{5} \le \dfrac{n_1}{n_2} \le 5$) are satisfied, then

$$t = \frac{\bar{A} - \bar{B}}{\sqrt{\dfrac{\Sigma(B - \bar{B})^2 + \Sigma(A - \bar{A})^2}{n_A + n_B - 2}\left(\dfrac{1}{n_A} + \dfrac{1}{n_B}\right)}} \qquad (6.16)^*$$

has approximately the Student t distribution with $df = n_A + n_B - 2$. Therefore the probability associated with the occurrence under H_0 of any value as extreme as an observed t may be determined by reference to Table B of the Appendix.

The reader should note that even though formula (6.16) is the ordinary t test, the test is not used in this case as a parametric statistical test, for the assumption that the populations are normally distributed with common variance is not necessary. However, its use requires not only that the two conditions mentioned above be met, but also that the scores represent measurement in at least an interval scale.

When n_1 and n_2 are large, another alternative to the randomization test is the Mann-Whitney U test, which may be regarded as a randomization test applied to the ranks of the observations and which thus constitutes a good approximation to the randomization test. It can be shown (Whitney, 1948) that there are situations under which the Mann-

* The barred symbols, for example, $\bar{B}$, stand for means.

Whitney U test is more powerful than the t test and thus is the better alternative.

Summary of procedure. These are the steps in the use of the randomization test for two independent samples:

1. Determine the number of possible outcomes in the region of rejection: $\alpha \begin{pmatrix} n_1 + n_2 \\ n_1 \end{pmatrix}$.

2. Specify as belonging to the region of rejection that number of the most extreme possible outcomes. The extremes are those which have the largest difference between ΣA and ΣB. For a one-tailed test, all of these are in the predicted direction. For a two-tailed test, half of the number are the most extreme possible outcomes in one direction and half are the most extreme possible outcomes in the other direction.

3. If the observed scores are one of the outcomes listed in the region of rejection, reject H_0 at the α level of significance.

For samples which are so large that the enumeration of the possible outcomes in the region of rejection is too tedious, formula (6.16) may be used as an approximation if the conditions for its use are met by the data. An alternative, which need not meet such conditions and thus may be more satisfactory, is the Mann-Whitney U test.

Power-Efficiency

Because it uses all the information in the samples, the randomization test for two independent samples has power-efficiency, in the sense defined, of 100 per cent.

References

The reader may find discussions of the randomization test for two independent samples in Moses (1952a), Pitman (1937a; 1937b; 1937c), Scheffé (1943), Smith (1953), and Welch (1937).

DISCUSSION

In this chapter we have presented eight statistical tests which are useful in testing for the "significance of the difference" between two independent samples. In his choice among these tests, the researcher may be aided by the discussion which follows, in which any unique advantages of the tests are pointed out and the contrasts among them are noted.

All the nonparametric tests for two independent samples test whether it is likely that the two independent samples came from the same population. But the various tests we have presented are more or less sensitive to different kinds of differences between samples. For example, if one wishes to test whether two samples represent populations which differ in location (central tendency), these are the tests which are most sensitive to such a difference and therefore should be chosen: the median test (or the Fisher test when N is small), the Mann-Whitney U test, the Kolmogorov-Smirnov two-sample test (for one-tailed tests), and the randomization test. On the other hand, if the researcher is interested in determining whether his two samples are from populations which differ in any respect at all, i.e., in location *or* dispersion *or* skewness, etc., he should choose one of these tests: the χ^2 test, the Kolmogorov-Smirnov test (two-tailed), or the Wald-Wolfowitz runs test. The remaining technique, the Moses test, is uniquely suitable for testing whether an experimental group is exhibiting extremist or defensive reactions in comparison to the reactions exhibited by an independent control group.

The choice among the tests which are sensitive to differences in location is determined by the kind of measurement achieved in the research and by the size of the samples. The most powerful test of location is the randomization test. However, this test can be used only when the sample sizes are small and when we have some confidence in the *numerical* nature of the measurement obtained. With larger samples or weaker measurement (ordinal measurement), the suggested alternative is the Mann-Whitney U test, which is almost as powerful as the randomization test. If the samples are very small, the Kolmogorov-Smirnov test is slightly more efficient than the U test. If the measurement is such that it is meaningful only to dichotomize the observations as above or below the combined median, then the median test is applicable. This test is not as powerful as the Mann-Whitney U test in guarding against differences in location, but it is more appropriate than the U test when the data are observations which cannot be completely ranked. If the combined sample sizes are very small, when applying the median test the researcher should make the analysis by the Fisher test.

The choice among the tests which are sensitive to all kinds of differences (the second group listed above) is predicated on the strength of the measurement obtained, the size of the two samples, and the relative power of the available tests. The χ^2 test is suitable for data which are in nominal or stronger scales. When the N is small and the data are in a 2×2 contingency table, the Fisher test should be used rather than χ^2. In many cases the χ^2 test may not make efficient use of all the information in the data. If the populations of scores are continuously distributed, we may choose either the Kolmogorov-Smirnov (two-tailed) test or the Wald-Wolfowitz runs test in preference to the χ^2 test. Of all tests for any kind of difference, the Kolmogorov-Smirnov test is the most powerful. If it is used with data which do not meet the assumption of continuity, it is still suitable but it operates more conservatively (Goodman, 1954), i.e., the obtained value of p in such cases will be slightly higher than it should be, and thus the probability of a Type II error will be slightly increased. If H_0 is rejected with such data, we can surely have confidence in the decision. The runs test also guards against all kinds of differences, but it is not as powerful as the Kolmogorov-Smirnov test.

Two points should be emphasized about the use of the second group of tests. First, if one is interested in testing the alternative hypothesis that the groups differ in central tendency, e.g., that one population has a larger median than the other, then one should use a test specifically designed to guard against differences in location—one of the tests in the first group listed above. Second, when one rejects H_0 on the basis of a test which guards against any kind of difference (one of the tests in the second group), one can then assert that the two groups are from different populations but one cannot say *in what specific way(s)* the populations differ.

CHAPTER 7

THE CASE OF k RELATED SAMPLES

In previous chapters we have presented statistical tests for (a) testing for significant differences between a single sample and some specified population, and (b) testing for significant differences between two samples, either related or independent. In this and the following chapters, procedures will be presented for testing for the significance of differences among three or more groups. That is, statistical tests will be presented for testing the null hypothesis that k (3 or more) samples have been drawn from the same population or from identical populations. This chapter will present tests for the case of k *related* samples; the following chapter will present tests for the case of k *independent* samples.

Circumstances sometimes require that we design an experiment so that more than two samples or conditions can be studied simultaneously. When three or more samples or conditions are to be compared in an experiment, it is necessary to use a statistical test which will indicate whether there is an *over-all* difference among the k samples or conditions before one picks out any pair of samples in order to test the significance of the difference between them.

If we wished to use a two-sample statistical test to test for differences among, say, 5 groups, we would need to compute, in order to compare each pair of samples, 10 statistical tests. (Five things taken 2 at a time $= \binom{5}{2} = 10$.) Such a procedure is not only tedious, but it may lead to fallacious conclusions as well because it capitalizes on chance. That is, suppose we wish to use a significance level of, say, $\alpha = .05$. Our hypothesis is that there is a difference among $k = 5$ samples. If we test that hypothesis by comparing each of the 5 samples with every other sample, using a two-sample test (which would require 10 comparisons in all), we are giving ourselves 10 chances rather than 1 chance to reject H_0. Now when we set .05 as our level of significance, we are taking the risk of rejecting H_0 erroneously (making the Type I error) 5 per cent of the time. But if we make 10 statistical tests of the same hypothesis, we increase the probability of the Type I error. It can be shown that, for 5 samples, the probability that a two-sample statistical test will find one or more "significant" differences, when $\alpha = .05$, is $p = .40$. That is, the *actual* significance level in such a procedure becomes $\alpha = .40$.

Cases have been reported in the research literature (McNemar, 1955, p. 234) in which an over-all test of five samples yields insignificant results (leads to the acceptance of H_0) but two-sample tests of the larger differences among the five samples yield significant findings. Such a posteriori selection tends to capitalize on chance, and therefore we can have no confidence in a decision involving k samples in which the analysis consisted only of testing two samples at a time.

It is only when an over-all test (a k-sample test) allows us to reject the null hypothesis that we are justified in employing a procedure for testing for significant differences between any two of the k samples. (For such a procedure, see Cochran, 1954; and Tukey, 1949.)

The parametric technique for testing whether several samples have come from identical populations is the analysis of variance or F test. The assumptions associated with the statistical model underlying the F test are these: that the scores or observations are independently drawn from normally distributed populations; that the populations all have the same variance; and that the means in the normally distributed populations are linear combinations of "effects" due to rows and columns, i.e., that the effects are additive. In addition, the F test requires at least interval measurement of the variables involved.

If a researcher finds such assumptions unrealistic for his data, if he finds that his scores do not meet the measurement requirement, or if he wishes to avoid making the assumptions in order to increase the generality of his findings, he may use one of the nonparametric statistical tests presented in this and the following chapter. In addition to avoiding the assumptions and requirements mentioned, these nonparametric k-sample tests have the further advantage of enabling data which are inherently only classificatory or in ranks to be examined for significance.

There are two basic designs for comparing k groups. In the first design, the k samples of equal size are *matched* according to some criterion or criteria which may affect the values of the observations. In some cases, the matching is achieved by comparing the same individuals or cases under all k conditions. Or each of N groups may be measured under all k conditions. For such designs, the statistical tests for k related samples (presented in this chapter) should be used. The second design involves k *independent* random samples, not necessarily of the same size, one sample from each population. For that design, the statistical tests for k independent samples (presented in Chap. 8) should be employed.

The above distinction is, of course, exactly that made in the parametric case. The first design is known as the two-way analysis of variance, sometimes called "the randomized blocks design."[1]   The second design is known as the one-way analysis of variance.

The distinction is similar to that we made between the case of two related samples (discussed in Chap. 5) and the case of two independent samples (discussed in Chap. 6).

This chapter will present nonparametric statistical tests which parallel the two-way analysis of variance.   We will present a test suitable for use with data measured in a nominal scale and another suitable for use with data measured in at least an ordinal scale.   At the conclusion of this chapter we shall compare and contrast these tests for k related samples, offering further guidance to the researcher in his selection of the test suitable for his data.

THE COCHRAN Q TEST
Function

The McNemar test for two related samples, presented in Chap. 5, can be extended for use in research having more than two samples.   This extension, the Cochran Q test for k related samples, provides a method for testing whether three or more matched sets of frequencies or proportions differ significantly among themselves.   The matching may be based on relevant characteristics of the different subjects, or on the fact that the same subjects are used under different conditions.   The Cochran test is particularly suitable when the data are in a nominal scale or are dichotomized ordinal information.

One may imagine a wide variety of research hypotheses for which the data might be analyzed by the Cochran test.   For example, one might test whether the various items on a test differ in difficulty by analyzing data consisting of pass-fail information on k items for N individuals.   In this design, the k groups are considered "matched" because each person answers all k items.

On the other hand, we might have only one item to be analyzed, and wish to compare the responses of N subjects under k different conditions. Here again the "matching" is achieved by having the same subjects in every group, but now the groups differ in that each is under a different condition.   This would test whether the k conditions have a significant effect on the subjects' responses to the item.   For example, one might ask each member of a panel of voters which of two candidates they prefer

[1] The term "randomized blocks" derives from agricultural experimentation, in which plots of land may be used as experimental units.   A "block" consists of adjacent plots of land, and it is assumed that plots of land adjacent to each other are more alike (i.e., are better matched) than are plots remote from each other.   The k treatments, for example, k varieties of fertilizer, or k varieties of seed, are assigned at random, one to each of the k plots in a block; this is done with independent random assignment in each block.

at $k = 5$ times during the election season: before the campaign, at the peak of Smith's campaign, at the peak of Miller's campaign, immediately before the balloting, and immediately after the results are announced. The Cochran test would determine whether these conditions have a significant effect on the voters' preferences between the candidates.

Again, we might compare the responses to one item from N sets having k matched persons in each set.   That is, we would have responses from k matched groups.

Method

If the data from researches like those exemplified above are arranged in a two-way table consisting of N rows and k columns, it is possible to test the null hypothesis that the proportion (or frequency) of responses of a particular kind is the same in each column, except for chance differences.   Cochran (1950) has shown that if the null hypothesis is true, i.e., if there is no difference in the probability of, say, "success" under each condition (which is to say that the "successes" and "failures" are randomly distributed in the rows and columns of the two-way table), then if the number of rows is not too small

$$Q = \frac{k(k-1) \sum_{j=1}^{k} (G_j - \bar{G})^2}{k \sum_{i=1}^{N} L_i - \sum_{i=1}^{N} L_i^2} \tag{7.1}$$

is distributed approximately as chi square with $df = k - 1$,
where G_j = total number of "successes" in jth column,
   $\bar{G}$ = mean of the G_j
   L_i = total number of "successes" in ith row
A formula which is equivalent to and easily derivable from (7.1) but which simplifies computation is

$$Q = \frac{(k-1) \left[k \sum_{j=1}^{k} G_j^2 - \left(\sum_{j=1}^{k} G_j \right)^2 \right]}{k \sum_{i=1}^{N} L_i - \sum_{i=1}^{N} L_i^2} \tag{7.2}$$

Inasmuch as the sampling distribution of Q is approximated by the chi-square distribution with $df = k - 1$, the probability associated with the occurrence under H_0 of values as large as an observed Q may be determined by reference to Table C of the Appendix.   If the observed value of Q, as computed from formula (7.2), is equal to or greater than that shown in Table C for a particular significance level and a particular value of $df = k - 1$, the implication is that the proportion (or frequency) of "successes" differs significantly among the various samples. That is, H_0 may be rejected at that particular level of significance.

Example

Suppose we were interested in the influence of interviewer friendliness upon housewives' responses in an opinion survey. We might train an interviewer to conduct three kinds of interviews: interview 1, showing interest, friendliness, and enthusiasm; interview 2, showing formality, reserve, and courtesy; and interview 3, showing disinterest, abruptness, and harsh formality. The interviewer would be assigned to visit 3 groups of 18 houses, and told to use interview 1 with one group, interview 2 with another, and interview 3 with the third. That is, we would have obtained 18 sets of housewives with 3 matched housewives (equated on relevant variables) in each set. For each set, the 3 members would randomly be assigned to the 3 conditions (types of interviews). Thus we would have 3 matched samples ($k = 3$) with 18 members in each ($N = 18$). We could then test whether the gross differences between the three styles of interviews influenced the number of "yes" responses given to a particular item by the three matched groups. Using artificial data, a test of this hypothesis follows.

i. *Null Hypothesis.* H_0: the probability of a "yes" response is the same for all three types of interviews. H_1: the probabilities of "yes" responses differ according to the style of the interview.

ii. *Statistical Test.* The Cochran Q test is chosen because the data are for more than two related groups ($k = 3$), and are dichotomized as "yes" and "no."

iii. *Significance Level.* Let $\alpha = .01$. $N = 18 =$ the number of cases in each of the k matched sets.

iv. *Sampling Distribution.* Under the null hypothesis, Q [as yielded by formula (7.1) or (7.2)] is distributed approximately as chi square with $df = k - 1$. That is, the probability associated with the occurrence under H_0 of any value as large as an observed value of Q may be determined by reference to Table C.

v. *Rejection Region.* The region of rejection consists of all values of Q which are so large that the probability associated with their occurrence under H_0 is equal to or less than $\alpha = .01$.

vi. *Decision.* In this artificial study, we will represent "yes" responses by 1's and "no" responses by 0's. The data of the study are given in Table 7.1. Also shown in that table are the values of L_i (the total number of yeses for each row) and the values of $L_i{}^2$. For example, in the first matched set all the housewives responded negatively, regardless of the interview style. Thus $L_1 = 0 + 0 + 0 = 0$, and thus $L_1{}^2 = 0^2 = 0$. In the second matched set of 3 housewives, the responses to interviews 1 and 2 were affirmative but the response to interview 3 was negative, so that $L_2 = 1 + 1 + 0 = 2$, and thus $L_2{}^2 = 2^2 = 4$.

As is the practice, the scores have been arranged in $k = 3$ columns and $N = 18$ rows.

TABLE 7.1. YES (1) AND NO (0) RESPONSES BY HOUSEWIVES UNDER
THREE TYPES OF INTERVIEWS
(Artificial data)

Set	Response to interview 1	Response to interview 2	Response to interview 3	L_i	$L_i{}^2$
1	0	0	0	0	0
2	1	1	0	2	4
3	0	1	0	1	1
4	0	0	0	0	0
5	1	0	0	1	1
6	1	1	0	2	4
7	1	1	0	2	4
8	0	1	0	1	1
9	1	0	0	1	1
10	0	0	0	0	0
11	1	1	1	3	9
12	1	1	1	3	9
13	1	1	0	2	4
14	1	1	0	2	4
15	1	1	0	2	4
16	1	1	1	3	9
17	1	1	0	2	4
18	1	1	0	2	4
	$G_1 = 13$	$G_2 = 13$	$G_3 = 3$	$\sum\limits_{i=1}^{18} L_i = 29$	$\sum\limits_{i=1}^{18} L_i{}^2 = 63$

We observe that $G_1 = 13 =$ the total number of yeses in response to interview 1. $G_2 = 13 =$ the total number of yeses in response to interview 2. And $G_3 = 3 =$ the total number of yeses in response to interview 3.

The total number of yeses in all three interviews $= \sum\limits_{j=1}^{3} G_j = 13$

$+ 13 + 3 = 29$. Observe that $\sum\limits_{i=1}^{18} L_i = 29$ also (the sum of the column of row totals). The sum of the squares of the row totals is $\sum_{i=1}^{18} L_i^2 = 63$, the sum of the final column.

By entering these values in formula (7.2) we have

$$Q = \frac{(k-1)\left[k \sum_{j=1}^{k} G_j^2 - \left(\sum_{j=1}^{k} G_j\right)^2\right]}{k \sum_{i=1}^{N} L_i - \sum_{i=1}^{N} L_i^2} \tag{7.2}$$

$$= \frac{(3-1)\{3[(13)^2 + (13)^2 + (3)^2] - (29)^2\}}{(3)(29) - 63}$$

$$= 16.7$$

Reference to Table C reveals that $Q \geq 16.7$ has probability of occurrence under H_0 of $p < .001$, when $df = k - 1 = 3 - 1 = 2$. This probability is smaller than the significance level, $\alpha = .01$. Thus the value of Q is in the region of rejection and therefore our decision is to reject H_0 in favor of H_1. On the basis of these artificial data, we conclude that the probabilities of "yes" responses are different under interviews 1, 2, and 3.

It should be noted that Q is distributed as chi square with $df = k - 1$ if the number of rows (the size of N) is not too small. However, Cochran makes no specific recommendations regarding the minimum size of N.

Summary of procedure. These are the steps in the use of the Cochran Q test:

1. For the dichotomous data, assign a score of 1 to each "success" and a score of 0 to each "failure."

2. Cast the scores in a $k \times N$ table, using k columns and N rows. $N =$ the number of cases in each of the k groups.

3. Determine the value of Q by substituting the observed values in formula (7.2).

4. The significance of the observed value of Q may be determined by reference to Table C, for Q is distributed approximately as chi square with $df = k - 1$. If the probability associated with the occurrence under H_0 of a value as large as the observed value of Q is equal to or less than α, reject H_0.

Power and Power-Efficiency

The power of the Cochran test is not known exactly. The notion of power-efficiency is meaningless when the Cochran test is used for nominal or naturally dichotomous data, for parametric tests are not applicable to such data. When the Cochran test is used with data that are not nominal or naturally dichotomous, it may be wasteful of information.

References

The reader may find discussions of the Cochran test in Cochran (1950) and McNemar (1955, pp. 232–233).

THE FRIEDMAN TWO-WAY ANALYSIS OF VARIANCE BY RANKS

Function

When the data from k matched samples are in at least an ordinal scale, the Friedman two-way analysis of variance by ranks is useful for testing the null hypothesis that the k samples have been drawn from the same population.

Since the k samples are matched, the number of cases is the same in each of the samples. The matching may be achieved by studying the same group of subjects under each of k conditions. Or the researcher may obtain several sets, each consisting of k matched subjects, and then randomly assign one subject in each set to the first condition, one subject in each set to the second condition, etc. For example, if one wished to study the differences in learning achieved under four teaching methods, one might obtain N sets of $k = 4$ pupils, each set consisting of children who are matched on the relevant variables (age, previous learning, intelligence, socioeconomic status, motivation, etc.), and then at random assign one child from each of the N sets to teaching method A, another from each set to B, another from each set to C, and the fourth to D.

Rationale and Method

For the Friedman test, the data are cast in a two-way table having N rows and k columns. The rows represent the various subjects or matched sets of subjects, and the columns represent the various conditions. If the scores of subjects serving under all conditions are under study, then each row gives the scores of one subject under the k conditions.

The data of the test are ranks. The scores in each *row* are ranked separately. That is, with k conditions being studied, the ranks in any row range from 1 to k. The Friedman test determines whether it is likely that the different *columns* of ranks (samples) came from the same population.

For example, suppose we wish to study the scores of 3 groups under 4 conditions. Here $k = 4$ and $N = 3$. Each group contains 4 matched subjects, one being assigned to each of the 4 conditions. Suppose our scores for this study were those given in Table 7.2. To perform the

TABLE 7.2. SCORES OF THREE MATCHED GROUPS UNDER FOUR CONDITIONS

	Conditions			
	I	II	III	IV
Group A	9	4	1	7
Group B	6	5	2	8
Group C	9	1	2	6

Friedman test on these data, we first rank the scores *in each row*. We may give the lowest score in each row the rank of 1, the next lowest score in each row the rank of 2, etc.[1] By doing this we obtain the data shown in Table 7.3. Observe that the ranks in each row of Table 7.3 range from 1 to $k = 4$.

TABLE 7.3. RANKS OF THREE MATCHED GROUPS UNDER FOUR CONDITIONS

	Conditions			
	I	II	III	IV
Group A	4	2	1	3
Group B	3	2	1	4
Group C	4	1	2	3
R_j	11	5	4	10

Now if the null hypothesis (that all the samples—columns—came from the same population) is in fact true, then the distribution of ranks *in each column* would be a matter of chance, and thus we would expect the ranks of 1, 2, 3, and 4 to appear in all columns with about equal frequency. This would indicate that for any group it is a matter of chance under which condition the highest score occurs and under which condition the lowest occurs, which would be the case if the conditions really did not differ.

If the subjects' scores were independent of the conditions, the set of ranks in each column would represent a random sample from the discontinuous rectangular distribution of 1, 2, 3, and 4, and the rank totals for the various columns would be about equal. If the subjects' scores were dependent on the conditions (i.e., if H_0 were false), then the rank totals would vary from one column to another. Inasmuch as the columns

[1] It is immaterial whether the ranking is from lowest to highest scores or from highest to lowest scores.

all contain an equal number of cases, an equivalent statement would be that under H_0 the mean ranks of the various columns would be about equal.

The Friedman test determines whether the rank totals (R_j) differ significantly. To make this test, we compute the value of a statistic which Friedman denotes as χ_r^2.

When the number of rows and/or columns is not too small, it can be shown (Friedman, 1937) that χ_r^2 is distributed approximately as chi square with $df = k - 1$, when

$$\chi_r^2 = \frac{12}{Nk(k+1)} \sum_{j=1}^{k} (R_j)^2 - 3N(k+1) \tag{7.3}$$

where N = number of rows
$\quad\ k$ = number of columns
$\quad R_j$ = sum of ranks in jth column

$\sum_{j=1}^{k}$ directs one to sum the squares of the sums of ranks over all k conditions

Inasmuch as the sampling distribution of χ_r^2 is approximated by the chi-square distribution with $df = k - 1$, the probability associated with the occurrence under H_0 of values as large as an observed χ_r^2 is shown in Table C of the Appendix. If the value of χ_r^2 as computed from formula (7.3) is equal to or larger than that given in Table C for a particular level of significance and a particular value of $df = k - 1$, the implication is that the sums of ranks (or, equivalently, the mean ranks, R_j/N) for the various columns differ significantly (which is to say that the size of the scores depends on the conditions under which the scores were obtained) and thus H_0 may be rejected at that level of significance.

Notice that χ_r^2 is distributed approximately as chi square with $df = k - 1$ only when the number of rows and/or columns is not too small. When the number of rows or columns is less than minimal, exact probability tables are available, and these should be used rather than Table C. Table N of the Appendix gives exact probabilities associated with values as large as an observed χ_r^2 for $k = 3$, $N = 2$ to 9, and for $k = 4$, $N = 2$ to 4. When N and k are larger than the values included in Table N, χ_r^2 may be considered to be distributed as chi square, and thus Table C may be used for testing H_0.

To illustrate the computation of χ_r^2 and the use of Table N, we may test for significance the data shown in Table 7.3. By referring to that table, the reader may see that the various sums of ranks, R_j, were 11, 5,

4, and 10. The number of conditions $= k = 4$. The number of rows $= N = 3$. We may compute the value of χ_r^2 for the data in Table 7.3 by substituting these values in formula (7.3):

$$\chi_r^2 = \frac{12}{Nk(k+1)} \sum_{j=1}^{k} (R_j)^2 - 3N(k+1) \tag{7.3}$$

$$= \frac{12}{(3)(4)(4+1)} [(11)^2 + (5)^2 + (4)^2 + (10)^2] - (3)(3)(4+1)$$

$$= 7.4$$

We may determine the probability of occurrence under H_0 of $\chi_r^2 \geq 7.4$ by turning to Table N_{II} which gives the exact probability associated with values as large as an observed χ_r^2 for $k = 4$. Table N shows that the probability associated with $\chi_r^2 \geq 7.4$ when $k = 4$ and $N = 3$ is $p = .033$. With these data, therefore, we could reject the null hypothesis that the four samples were drawn from the same population with respect to location (mean ranks) at the .033 level of significance.

Example for N and k Large

In a study of the effect of three different patterns of reinforcement upon extent of discrimination learning in rats,[1] three matched samples ($k = 3$) of 18 rats ($N = 18$) were trained under three patterns of reinforcement. Matching was achieved by the use of 18 sets of littermates, 3 in each set. Although all the 54 rats received the same quantity of reinforcement (reward), the patterning of the administration of reinforcement was different for each of the groups. One group was trained with 100 per cent reinforcement (RR), a matched group was trained with partial reinforcement in which each sequence of trials ended with an unreinforced trial (RU), and the third matched group was trained with partial reinforcement in which each sequence of trials ended with a reinforced trial (UR).

After this training, the extent of learning was measured by the speed at which the various rats learned an "opposing" habit: whereas they had been trained to run to white, the rats now were rewarded for running to black. The better the initial learning, the slower this transfer of learning should be. The prediction was that the different patterns of reinforcement used would result in differential learning as exhibited by ability to transfer.

---

[1] Grosslight, J. H., and Radlow, R. 1956. Studies in partial reinforcement: I. Patterning effect of the nonreinforcement-reinforcement sequence in a discrimination situation. *J. Comp. Physiol. Psychol.*, in press.

i. *Null Hypothesis.* H_0: the different patterns of reinforcement have no differential effect. H_1: the different patterns of reinforcement have a differential effect.

ii. *Statistical Test.* Because number of errors in transfer of learning is probably not an interval measure of strength of original learning, the nonparametric two-way analysis of variance was chosen rather than the parametric. Moreover, the use of the parametric analysis of variance was also precluded because the scores exhibited possible lack of homogeneity of variance and thus the data suggested that one of the basic assumptions of the F test was probably untenable.

iii. *Significance Level.* Let $\alpha = .05$. $N = 18 = $ the number of rats in each of the 3 matched groups.

iv. *Sampling Distribution.* As computed by formula (7.3), χ_r^2 is distributed approximately as chi square with $df = k - 1$ when N and/or k are large. Thus the probability associated with the occurrence under H_0 of a value as large as the observed value of χ_r^2 may be determined by reference to Table C.

v. *Rejection Region.* The region of rejection consists of all values of χ_r^2 which are so large that the probability associated with their occurrence under H_0 is equal to or less than $\alpha = .05$.

vi. *Decision.* The number of errors committed by each rat in the transfer of learning situation was determined, and these scores were ranked for each of the 18 sets of 3 matched rats. These ranks are given in Table 7.4.

Observe that the sum of ranks for the RR group is 39.5, the sum of ranks for the RU group is 42.5, and the sum of ranks for the UR group is 26.0. A low rank signifies a high number of errors in transfer, i.e., signifies strong original learning. We may compute the value of χ_r^2 by substituting our observed values in formula (7.3):

$$\chi_r^2 = \frac{12}{Nk(k+1)} \sum_{j=1}^{k} (R_j)^2 - 3N(k+1) \qquad (7.3)$$

$$= \frac{12}{(18)(3)(3+1)} [(39.5)^2 + (42.5)^2 + (26.0)^2] - (3)(18)(3+1)$$

$$= 8.4$$

Reference to Table C indicates that $\chi_r^2 = 8.4$ when $df = k - 1 = 3 - 1 = 2$ is significant at between the .02 and .01 levels. Since $p < .02$ is less than our previously established significance level of $\alpha = .05$, the decision is to reject H_0. The conclusion is that rats' scores on transfer of learning depend on the pattern of reinforcement in the original learning trials.

TABLE 7.4. RANKS OF EIGHTEEN MATCHED GROUPS IN TRANSFER
AFTER TRAINING UNDER THREE CONDITIONS OF REINFORCEMENT

Group	Type of reinforcement		
	RR	RU	UR
1	1	3	2
2	2	3	1
3	1	3	2
4	1	2	3
5	3	1	2
6	2	3	1
7	3	2	1
8	1	3	2
9	3	1	2
10	3	1	2
11	2	3	1
12	2	3	1
13	3	2	1
14	2	3	1
15	2.5*	2.5*	1
16	3	2	1
17	3	2	1
18	2	3	1
R_j	39.5	42.5	26.0

* In group 15, the RR and the RU animals obtained equal scores and thus were tied for ranks 2 and 3. Both were given ranks 2.5, the average of the tied ranks. Friedman (1937, p. 681) states that the substitution of the average rank for tied values does not affect the validity of the χ_r^2 test.

Summary of procedure.   These are the steps in the use of the Friedman two-way analysis of variance by ranks:

1. Cast the scores in a two-way table having k columns (conditions) and N rows (subjects or groups).

2. Rank the scores in each row from 1 to k.

3. Determine the sum of the ranks in each column: R_j.

4. Compute the value of χ_r^2, using formula (7.3).

5. The method for determining the probability of occurrence under H_0 associated with the observed value of χ_r^2 depends on the sizes of N and k:

   a. Table N gives exact probabilities associated with values as large as an observed χ_r^2 for $k = 3$, $N = 2$ to 9, and for $k = 4$, $N = 2$ to 4.

   b. For N and/or k larger than those shown in Table N, the associated probability may be determined by reference to the chi-square distribution (given in Table C) with $df = k - 1$.

6. If the probability yielded by the appropriate method in step 5 is equal to or less than α, reject H_0.

Power

The exact power of the χ_r^2 test is not reported in the literature. However, Friedman (1937, p. 686) has reported the results of 56 independent analyses of data which were suitable for analysis by the parametric F test and which were analyzed by both that test and by the nonparametric χ_r^2 test. The results give a good idea of the efficiency of the χ_r^2 test as compared to the most powerful k-sample parametric test (under these conditions): the F test. They are given in Table 7.5. The reader can

TABLE 7.5. COMPARISON OF RESULTS OF THE F TEST AND THE χ_r^2 TEST ON 56 SETS OF DATA WHICH MET THE ASSUMPTIONS AND REQUIREMENTS OF THE F TEST

Number of χ_r^2's with probability	Number of F's with probability			Total
	Greater than .05	Between .05 and .01	Less than .01	
Greater than .05	28	2	0	30
Between .05 and .01	4	1	4	9
Less than .01	0	1	16	17
Total	32	4	20	56

see from the information in that table that it would be difficult or even impossible to say which is the more powerful test. In no case did one of the tests yield a probability of less than .01, while the other yielded a probability greater than .05. In 45 of the 56 cases, the probability levels yielded by the two tests were essentially the same. For the 56 sets of data, the χ_r^2 test would have rejected H_0 26 times at $\alpha = .05$, while the F test would have rejected H_0 24 times at that significance level.

References

The reader may find discussions of the Friedman two-way analysis of variance by ranks in Friedman (1937; 1940) and in Kendall (1939: 1948a, chaps. 6, 7).

DISCUSSION

Two nonparametric statistical tests for testing H_0 in the case of k related samples were presented in this chapter. The first, the Cochran Q test, is useful when the measurement of the variable under study is in a nominal or dichotomized ordinal scale. This test determines whether it is likely that the k related samples could have come from the same population with respect to proportion or frequency of "successes" in the various samples. That is, it is an over-all test of whether the k samples exhibit significantly different frequencies of "successes."

The second statistical test presented, the Friedman χ_r^2 test, is useful when the measurement of the variable is in at least an ordinal scale. It tests whether the k related samples could probably have come from the same population with respect to mean ranks. That is, it is an over-all test of whether the size of the scores depends on the conditions under which they were yielded.

Very little is known about the power of either test. However, the empirical study by Friedman which was cited earlier has shown very favorable results for the χ_r^2 test as compared with the most powerful parametric test, the F test. This would suggest that the Friedman test should be used in preference to the Cochran test whenever the data are appropriate (i.e., whenever the scores are in at least an ordinal scale). The χ_r^2 test has the further advantage of having tables of exact probabilities for very small samples, whereas the Cochran test should not be used when N (the number of rows) is too small.

CHAPTER 8

THE CASE OF k INDEPENDENT SAMPLES

In the analysis of research data, the investigator often needs to decide whether several independent samples should be regarded as having come from the same population. Sample values almost always differ somewhat, and the problem is to determine whether the observed sample differences signify differences among populations or whether they are merely the chance variations that are to be expected among random samples from the same population.

In this chapter, procedures will be presented for testing for the significance of differences among three or more independent groups or samples. That is, statistical techniques will be presented for testing the null hypothesis that k independent samples have been drawn from the same population or from k identical populations.

In the introduction to Chap. 7, we attempted to distinguish between two sorts of k-sample tests. The first sort of test is useful for analyzing data from k *matched* samples, and two nonparametric tests of this sort were presented in Chap. 7. The second kind of k-sample test is useful for analyzing data from k *independent* samples. Such tests will be presented in this chapter.

The usual parametric technique for testing whether several independent samples have come from the same population is the one-way analysis of variance or F test. The assumptions associated with the statistical model underlying the F test are that the observations are independently drawn from normally distributed populations, all of which have the same variance. The measurement requirement of the F test is that the research must achieve at least interval measurement of the variable involved.

If a researcher finds such assumptions are unrealistic for his data, or if his measurement is weaker than interval scaling, or if he wishes to avoid making the restrictive assumptions of the F test and thus to increase the generality of his findings, he may use one of the nonparametric statistical tests for k independent samples which are presented in this chapter. These nonparametric tests have the further advantage of enabling data

174

which are inherently only classificatory (in a nominal scale) or in ranks (in an ordinal scale) to be examined for significance.

We shall present three nonparametric tests for the case of k independent samples, and shall conclude the chapter with a discussion of the comparative uses of these tests.

THE χ^2 TEST FOR k INDEPENDENT SAMPLES

Function

When frequencies in discrete categories (either nominal or ordinal) constitute the data of research, the χ^2 test may be used to determine the significance of the differences among k independent groups. The χ^2 test for k independent samples is a straightforward extension of the χ^2 test for two independent samples, which is presented in Chap. 6. In general, the test is the same for both two and k independent samples.

Method

The method of computing the χ^2 test for independent samples will be presented briefly here, together with an example of the application of the test. The reader will find a fuller discussion of this test in Chap. 6.

To apply the χ^2 test, one first arranges the frequencies in a $k \times r$ table. The null hypothesis is that the k samples of frequencies or proportions have come from the same population or from identical populations. This hypothesis, that the k samples do not differ among themselves, may be tested by applying formula (6.3):

$$\chi^2 = \sum_{i=1}^{r} \sum_{j=1}^{k} \frac{(O_{ij} - E_{ij})^2}{E_{ij}} \qquad (6.3)$$

where O_{ij} = observed number of cases categorized in ith row of jth column

E_{ij} = number of cases expected under H_0 to be categorized in ith row of jth column, as determined by method presented on page 105.

$\sum_{i=1}^{r} \sum_{j=1}^{k}$ directs one to sum over all cells

Under H_0, the sampling distribution of χ^2 as computed from formula (6.3) can be shown to be approximated by a chi-square distribution with $df = (k - 1)(r - 1)$, where k = the number of columns and r = the number of rows. Thus, the probability associated with the occurrence of values as large as an observed χ^2 is given in Table C of the Appendix. If an observed value of χ^2 is equal to or larger than that given in Table C for a particular level of significance and for $df = (k - 1)(r - 1)$, then H_0 may be rejected at that level of significance.

Example

In an investigation of the nature and consequences of social stratification in a small Middle Western community,[1] Hollingshead found
that members of the community divided themselves into five social
classes: I, II, III, IV, and V. His research centered on the correlates of this stratification among the youth of the community. One
of his predictions was that adolescents in the different social classes
would enroll in different curriculums (college preparatory, general,
commercial) at the Elmtown high school. He tested this by
identifying the social class membership of 390 high school students
and determining the curriculum enrollment of each.

i. *Null Hypothesis.* H_0: the proportion of students enrolled in the
three alternative curriculums is the same in all classes. H_1: the
proportion of students enrolled in the three alternative curriculums
differs from class to class.

ii. *Statistical Test.* Since the groups under study are independent
and number more than 2, a statistical test for k independent samples
is called for. Since the data are in discrete categories, the χ^2 test is
an appropriate one.

iii. *Significance Level.* Let $\alpha = .01$. $N = 390$, the number of
students whose enrollment and class status were observed.

iv. *Sampling Distribution.* Under the null hypothesis, χ^2 as
computed from formula (6.3) is distributed approximately as chi
square with $df = (k - 1)(r - 1)$. The probability associated with
the occurrence under H_0 of values as large as an observed value of
χ^2 is shown in Table C.

v. *Rejection Region.* The region of rejection consists of all values
of χ^2 which are so large that the probability associated with their
occurrence under H_0 is equal to or less than $\alpha = .01$.

vi. *Decision.* Table 8.1 gives the curricular enrollment of the
390 high school students in Elmtown who were studied by Hollingshead. Social classes I and II were grouped together by Hollingshead because of the small number of youths belonging to these two
classes, particularly to class I. Table 8.1 also shows in italics the
number of youths who might be expected under H_0 to have enrolled
in each of the three curriculums, i.e., the expected enrollments if
there were really no difference in enrollment among the various social
classes. (These expected values were determined from the marginal
totals by the method presented on page 105.) For example, whereas
23 of the class I and II students were enrolled in the college preparatory curriculum, under H_0 we would expect only 7.3 to have enrolled

[1] Hollingshead, A. B. 1949. *Elmtown's youth: The impact of social classes on
adolescents.* New York: Wiley.

in that curriculum.   And whereas only 1 of the class I and II students enrolled in the commercial curriculum, under H_0 we would expect 9.1 to have enrolled in that curriculum.   Of the 26 class V youths, only 2 were enrolled in the college preparatory curriculum, whereas under H_0 we would expect to find 5.4 in that curriculum.

TABLE 8.1. FREQUENCY OF ENROLLMENT OF ELMTOWN YOUTHS FROM FIVE SOCIAL CLASSES IN THREE ALTERNATIVE HIGH SCHOOL CURRICULUMS*

Curriculum	Class				Total
	I and II	III	IV	V	
College preparatory	*7.3* 23	*30.3* 40	*38.0* 16	*5.4* 2	81
General	*18.6* 11	*77.5* 75	*97.1* 107	*13.8* 14	207
Commercial	*9.1* 1	*38.2* 31	*47.9* 60	*6.8* 10	102
Total	35	146	183	26	390

* Adapted from Table X of Hollingshead, A. B. 1949. *Elmtown's youth.* New York: Wiley, p. 462, with the kind permission of John Wiley & Sons, Inc.

The size of χ^2 reflects the magnitude of the discrepancy between the observed and the expected values in each of the cells. We may compute χ^2 for the values in Table 8.1 by the application of formula (6.3):

$$\chi^2 = \sum_{i=1}^{r} \sum_{j=1}^{k} \frac{(O_{ij} - E_{ij})^2}{E_{ij}} \tag{6.3}$$

$$= \frac{(23 - 7.3)^2}{7.3} + \frac{(40 - 30.3)^2}{30.3} + \frac{(16 - 38.0)^2}{38.0} + \frac{(2 - 5.4)^2}{5.4}$$

$$+ \frac{(11 - 18.6)^2}{18.6} + \frac{(75 - 77.5)^2}{77.5} + \frac{(107 - 97.1)^2}{97.1}$$

$$+ \frac{(14 - 13.8)^2}{13.8} + \frac{(1 - 9.1)^2}{9.1} + \frac{(31 - 38.2)^2}{38.2} + \frac{(60 - 47.9)^2}{47.9}$$

$$+ \frac{(10 - 6.8)^2}{6.8}$$

$$= 33.8 + 3.1 + 12.7 + 2.1 + 3.1 + .08 + 1.0 + .003 + 7.3$$
$$+ 1.4 + 3.1 + 1.5$$

$$= 69.2$$

We observe that for the data in Table 8.1, $\chi^2 = 69.2$ with

$$df = (k - 1)(r - 1) = (4 - 1)(3 - 1) = 6$$

Reference to Table C reveals that such a value of χ^2 is significant far beyond the .001 level. Since $p < .001$ is less than our previously set level of significance, $\alpha = .01$, our decision is to reject H_0. We conclude that high school students' curricular enrollment is not independent of social class membership among Elmtown's youth.

Summary of procedure. These are the steps in the use of the χ^2 test for k independent samples:

1. Cast the observed frequencies in a $k \times r$ contingency table, using the k columns for the groups.

2. Determine the expected frequency under H_0 for each cell by finding the product of the marginal totals common to the cell and dividing this product by N. (N is the sum of each group of marginal totals. It represents the total number of *independent* observations. Inflated N's invalidate the test.)

3. Compute χ^2 by using Formula (6.3). Determine

$$df = (k - 1)(r - 1)$$

4. Determine the significance of the observed value of χ^2 by reference to Table C. If the probability given for the observed value of χ^2 for the observed value of df is equal to or smaller than α, reject H_0 in favor of H_1.

When to Use the χ^2 Test

The χ^2 test requires that the expected frequencies (E_{ij}'s) in each cell should not be too small. When this requirement is violated, the results of the test are meaningless. Cochran (1954) recommends that for χ^2 tests with df larger than 1 (that is, when either k or r is larger than 2), fewer than 20 per cent of the cells should have an expected frequency of less than 5, and no cell should have an expected frequency of less than 1.

If these requirements are not met by the data in the form in which they were originally collected, the researcher must combine adjacent categories so as to increase the E_{ij}'s in the various cells. Only after he has combined the categories so that fewer than 20 per cent of the cells have expected frequencies of less than 5 and no cell has an expected frequency of less than 1 can the researcher meaningfully apply the χ^2 test. Of course he will be limited in his combining by the nature of his data. That is, the results of the statistical test may not be interpretable if the combining of categories has been capricious. The adjacent categories which are combined must have some common property or mutual identity if interpretation of the outcome of the test after the combining is to be possible. The researcher will guard against the necessity of combining categories if he uses a sufficiently large N in his research.

χ^2 tests are insensitive to the effects of order when $df > 1$. Thus when a hypothesis takes order into account, χ^2 may not be the best test. Cochran (1954) has presented methods that strengthen the common χ^2 tests when H_0 is tested against specific alternatives.

Power

There is usually no clear alternative to the χ^2 test when it is used, and thus the exact power of the χ^2 test usually cannot be computed. However, Cochran (1952, pp. 323–324) has shown that the limiting power distribution of χ^2 tends to 1 as N becomes large.

References

For other discussions of the χ^2 test, the reader is referred to Cochran (1952; 1954), Dixon and Massey (1951, chap. 13), Edwards (1954, chap. 18), Lewis and Burke (1949), McNemar (1955, chap. 13), and Walker and Lev (1953, chap. 4).

THE EXTENSION OF THE MEDIAN TEST
Function

The extension of the median test determines whether k independent groups (not necessarily of equal size) have been drawn from the same population or from populations with equal medians. It is useful when the variable under study has been measured in at least an ordinal scale.

Method

To apply the extension of the median test, we first determine the median score for the combined k samples of scores, i.e., we find the common median for all scores in the k groups. We then replace each score by a plus if the score is larger than the common median and by a minus if it is smaller than the common median. (If it happens that one or more scores fall at the common median, then the scores may be dichotomized by assigning a plus to those scores which exceed the common median and a minus to those which fall at the median or below.)

We may cast the resulting dichotomous sets of scores into a $k \times 2$ table, with the numbers in the body of the table representing the frequencies of pluses and minuses in each of the k groups. Table 8.3, shown below, is an example of such a table.

To test the null hypothesis that the k samples have come from the same population with respect to medians, we compute the value of χ^2 from formula (6.3):

$$\chi^2 = \sum_{i=1}^{r} \sum_{j=1}^{k} \frac{(O_{ij} - E_{ij})^2}{E_{ij}} \tag{6.3}$$

where O_{ij} = observed number of cases categorized in ith row of jth column

E_{ij} = number of cases expected under H_0 to be categorized in ith row of jth column

$\sum_{i=1}^{r} \sum_{j=1}^{k}$ directs one to sum over all cells

It can be shown that the sampling distribution under H_0 of χ^2 as computed from formula (6.3) is approximated by the chi-square distribution with $df = (k-1)(r-1)$, where k = the number of columns and r = the number of rows. In the median test, $r = 2$, and thus

$$df = (k-1)(r-1) = (k-1)(2-1) = (k-1)$$

The probability associated with the occurrence under H_0 of values as large as an observed value of χ^2 are given in Table C of the Appendix. If the observed value of χ^2 is equal to or larger than that given in Table C for the previously set level of significance and for the observed value of $df = k - 1$, then H_0 may be rejected at that level of significance.

If it is possible to dichotomize the scores exactly at the common median, then each E_{ij} is one-half of the marginal total for its column. When the scores are dichotomized as those which do and do not exceed the common median, the method for finding expected frequencies which is presented on page 105 should be used.

Once the data have been categorized as pluses and minuses with respect to the common median, and the resulting frequencies have been cast in a $k \times 2$ table, the computation procedures for this test are exactly the same as those for the χ^2 test for k independent samples, presented in the previous section of this chapter. They will be illustrated in the example which follows.

Example

Suppose an educational researcher wishes to study the influence of amount of education upon mothers' degree of interest in their children's schooling. He takes the highest school grade completed by each mother as an index to her amount of education; as an index to degree of interest in the child's schooling he takes the number of voluntary visits which each mother makes to the school during one school year, e.g., to class plays, to parent meetings, to self-initiated conferences with teachers and administrators, etc. By drawing every tenth name from the file of the names of the 440 children enrolled in the school, he obtains the names of 44 mothers, who constitute his sample. His hypothesis is that mothers' numbers of visits will vary according to the number of years the mothers have completed in school.

i. *Null Hypothesis.* H_0: there is no difference in frequency of school visits among mothers with different amounts of education, i.e., frequency of maternal visits to school is independent of amount of maternal education. H_1: the frequency of school visits by mothers differs for varying amounts of maternal education.

ii. *Statistical Test.* Since the groups of mothers of various educational levels are independent of each other and since several groups are anticipated, a significance test for k independent samples is in order. Since number of years of school constitutes at best an ordinal measure of degree of education, and since number of visits to school constitutes at best an ordinal measure of degree of interest in one's child's schooling, the extension of the median test is appropriate for testing the hypothesis concerning differences in central tendencies.

iii. *Significance Level.* Let $\alpha = .05$. $N = 44 =$ the number of mothers in the sample.

iv. *Sampling Distribution.* Under the null hypothesis, χ^2 as computed from formula (6.3) is distributed approximately as chi square with $df = k - 1$ when $r = 2$. The probability associated with the occurrence under H_0 of values as large as an observed χ^2 is shown in Table C.

v. *Rejection Region.* The region of rejection consists of all values of χ^2 which are so large that the probability associated with their occurrence under H_0 is equal to or less than $\alpha = .05$.

vi. *Decision.* In our fictitious example, the researcher collects the data presented in Table 8.2. The common median for these 44 scores is 2.5. That is, half of the mothers visited the school two or fewer times during the school year, and half visited three or more times. If we split each group of scores at that combined median, we obtain the data shown in Table 8.3, which gives the number of mothers in each educational level who fall above or below the common median in number of visits to school. Of those mothers whose education was limited to 8 years, for example, five visited the school three or more times and five visited two or fewer times. Of those mothers who had attended some years of college, three visited the school three or more times, and one visited two or fewer times.

Given in italics in Table 8.3 are the expected number of visits of each group under H_0. Observe that, with the scores dichotomized

TABLE 8.2. NUMBER OF VISITS TO SCHOOL BY MOTHERS OF
VARIOUS EDUCATIONAL LEVELS
(Artificial data)

Education completed by mother					
Elementary school (8th grade)	10th grade	High school (12th grade)	Some college	College graduate	Graduate school
4	2	2	9	2	2
3	4	0	4	4	6
0	1	4	2	5	
7	6	3	3	2	
1	3	8			
2	0	0			
0	2	5			
3	5	2			
5	1	1			
1	2	7			
	1	6			
		5			
		1			

TABLE 8.3. VISITS TO SCHOOL BY MOTHERS OF VARIOUS EDUCATIONAL LEVELS
(Artificial data)

	Education completed by mother						
	Elementary school (8th grade)	10th grade	High school (12th grade)	Some college	College graduate	Graduate school	Total
No. of mothers whose visits were more frequent than common median no. of visits	*5* 5	*5.5* 4	*6.5* 7	*2* 3	*2* 2	*1* 1	22
No. of mothers whose visits were less frequent than common median no. of visits	*5* 5	*5.5* 7	*6.5* 6	*2* 1	*2* 2	*1* 1	22
Total	10	11	13	4	4	2	44

exactly at the median, the expected frequency in each cell is just one-half of the total for the column in which the cell is located. Examining the data, the reader will notice that in this form the data are not amenable to a χ^2 analysis, for more than 20 per cent of the cells have expected frequencies of less than 5. (See the discussion of when to use the χ^2 test, on pages 178 and 179.) Observe that those categories with the unacceptably small expected frequencies all concern mothers who have attended college for various amounts of time: those who had some college, those who graduated from college, and those who attended graduate school. We may combine the three categories into one: College (one or more years). By doing so, we obtain the data shown in Table 8.4. Observe that in this

TABLE 8.4. VISITS TO SCHOOL BY MOTHERS OF VARIOUS EDUCATIONAL LEVELS
(Artificial data)

	Education completed by mother				
	Elementary school (8th grade)	10th grade	High school (12th grade)	College (one or more years)	Total
No. of mothers whose visits were more frequent than common median no. of visits	*5* 5	*5.5* 4	*6.5* 7	*5* 6	22
No. of mothers whose visits were less frequent than common median no. of visits	*5* 5	*5.5* 7	*6.5* 6	*5* 4	22
Total	10	11	13	10	44

table we have data which *are* amenable to a χ^2 analysis.

We may compute the observed value of χ^2 by substituting the data in Table 8.4 into formula (6.3):

$$\chi^2 = \sum_{i=1}^{r} \sum_{j=1}^{k} \frac{(O_{ij} - E_{ij})^2}{E_{ij}} \tag{6.3}$$

$$= \frac{(5 - 5)^2}{5} + \frac{(4 - 5.5)^2}{5.5} + \frac{(7 - 6.5)^2}{6.5} + \frac{(6 - 5)^2}{5}$$

$$+ \frac{(5 - 5)^2}{5} + \frac{(7 - 5.5)^2}{5.5} + \frac{(6 - 6.5)^2}{6.5} + \frac{(4 - 5)^2}{5}$$

$$= 0 + .409 + .0385 + .2 + 0 + .409 + .0385 + .2$$

$$= 1.295$$

By this computation we determine that $\chi^2 = 1.295$, and we know that $df = k - 1 = 4 - 1 = 3$. Reference to Table C reveals that under H_0 a χ^2 equal to or greater than 1.295 for $df = 3$ has probability of occurrence between .80 and .70. Since this p is larger than our previously set level of significance, $\alpha = .05$, our decision must be that on the basis of these (fictitious) data, we cannot reject the null hypothesis that the number of school visits made by mothers is independent of amount of maternal education.

Summary of procedure.   These are the steps in the use of the extension of the median test:

1. Determine the common median of the scores in the k groups.

2. Assign pluses to all scores above that median and minuses to all scores below, thereby splitting each of the k groups of scores at the combined median.   Cast the resulting frequencies in a $k \times 2$ table.

3. Using the data in that table, compute the value of χ^2 as given by formula (6.3).   Determine $df = k - 1$.

4. Determine the significance of the observed value of χ^2 by reference to Table C.   If the associated probability given for values as large as the observed value of χ^2 is equal to or smaller than α, reject H_0 in favor of H_1.

As we have mentioned, the extension of the median test is in essence a χ^2 test for k samples.   For information concerning the conditions under which the test may properly be used, and the power of the test, the reader is referred to discussions of these topics on pages 178 and 179

References

Discussions relevant to this test are contained in Cochran (1954) and Mood (1950, pp. 398–399).

THE KRUSKAL-WALLIS ONE-WAY ANALYSIS OF VARIANCE BY RANKS

Function

The Kruskal-Wallis one-way analysis of variance by ranks is an extremely useful test for deciding whether k independent samples are from different populations.   Sample values almost invariably differ somewhat, and the question is whether the differences among the samples signify genuine population differences or whether they represent merely chance variations such as are to be expected among several random samples from the same population.   The Kruskal-Wallis technique tests the null hypothesis that the k samples come from the same population or from identical populations with respect to averages.   The test assumes that the variable under study has an underlying continuous distribution. It requires at least ordinal measurement of that variable.

Rationale and Method

In the computation of the Kruskal-Wallis test, each of the N observations are replaced by ranks. That is, all of the scores from all of the k samples combined are ranked in a single series. The smallest score is replaced by rank 1, the next to smallest by rank 2, and the largest by rank N. N = the total number of independent observations in the k samples.

When this has been done, the sum of the ranks in each sample (column) is found. The Kruskal-Wallis test determines whether these sums of ranks are so disparate that they are not likely to have come from samples which were all drawn from the same population.

It can be shown that if the k samples actually are from the same population or from identical populations, that is, if H_0 is true, then H [the statistic used in the Kruskal-Wallis test and defined by formula (8.1) below] is distributed as chi square with $df = k - 1$, provided that the sizes of the various k samples are not too small. That is,

$$H = \frac{12}{N(N + 1)} \sum_{j=1}^{k} \frac{R_j{}^2}{n_j} - 3(N + 1) \tag{8.1}$$

where k = number of samples

n_j = number of cases in jth sample

$N = \Sigma n_j$, the number of cases in all samples combined

R_j = sum of ranks in jth sample (column)

$\sum_{j=1}^{k}$ directs one to sum over the k samples (columns)

is distributed approximately as chi square with $df = k - 1$, for sample sizes (n_j's) sufficiently large.

When there are more than 5 cases in the various groups, that is, $n_j > 5$, the probability associated with the occurrence under H_0 of values as large as an observed H may be determined by reference to Table C of the Appendix. If the observed value of H is equal to or larger than the value of chi square given in Table C for the previously set level of significance and for the observed value of $df = k - 1$, then H_0 may be rejected at that level of significance.

When $k = 3$ and the number of cases in each of the 3 samples is 5 or fewer, the chi-square approximation to the sampling distribution of H is not sufficiently close. For such cases, exact probabilities have been tabled from formula (8.1). These are presented in Table O of the Appendix. The first column in that table gives the number of cases in the 3 samples, i.e., gives various possible values of n_1, n_2, and n_3. The second gives various values of H, as computed from formula (8.1). The third gives the probability associated with the occurrence under H_0 of values as large as an observed H.

For example, if $H \geq 5.8333$ when the three samples contain 4, 3, and 1 cases, Table O show that the null hypothesis may be rejected at the .021 level of significance.

Example for Small Samples

Suppose an educational researcher wishes to test the hypothesis that school administrators are typically more authoritarian than classroom teachers. He knows, however, that his data for testing this hypothesis may be contaminated by the fact that many classroom teachers are administration-oriented in their professional aspirations. That is, many teachers take administrators as a reference group. To avoid this contamination, he plans to divide his 14 subjects into 3 groups: teaching-oriented teachers (classroom teachers who wish to remain in a teaching position), administration-oriented teachers (classroom teachers who aspire to become administrators), and administrators. He administers the F scale[1] (a measure of authoritarianism) to each of the 14 subjects. His hypothesis is that the three groups will differ with respect to averages on the F scale.

i. *Null Hypothesis.* H_0: there is no difference among the average F scores of teaching-oriented teachers, administration-oriented teachers, and administrators. H_1: the three groups of educators are not the same in their average F scores.[2]

ii. *Statistical Test.* Since three independent groups are under study, a test for k independent samples is called for. Since F-scale scores may be considered to represent at least ordinal measurement of authoritarianism, the Kruskal-Wallis test is appropriate.

iii. *Significance Level.* Let $\alpha = .05$. $N = 14 = $ the total number of educators studied. $n_1 = 5 = $ the number of teaching-oriented teachers. $n_2 = 5 = $ the number of administration-oriented teachers. $n_3 = 4 = $ the number of administrators.

iv. *Sampling Distribution.* For $k = 3$ and n_j's small, Table O gives the probability associated with the occurrence under H_0 for values as large as an observed H.

---

[1] Presented in Adorno, T. W., Frenkel-Brunswik, Else, Levinson, D. J., and Sanford, R. N. 1950. *The authoritarian personality.* New York: Harper.

[2] If X stands for the score of a teaching-oriented teacher, Y stands for the score of an administration-oriented teacher, and Z stands for the score of an administrator, then H_0, more properly stated, is that $p(X > Y) = p(X > Z) = p(Y > Z) = \frac{1}{2}$. H_1 would then call for inequality at least once.

v. *Rejection Region.* The region of rejection consists of all values of H which are so large that the probability associated with their occurrence under H_0 is equal to or less than $\alpha = .05$.

vi. *Decision.* For this fictitious study, the F scores for the various educators are shown in Table 8.5. If we rank these 14 F scores

TABLE 8.5. AUTHORITARIANISM SCORES OF THREE GROUPS OF EDUCATORS
(Artificial data)

Teaching-oriented teachers	Administration-oriented teachers	Administrators
96	82	115
128	124	149
83	132	166
61	135	147
101	109	

from lowest to highest, we obtain the ranks shown in Table 8.6. These ranks are summed for the three groups to obtain $R_1 = 22$, $R_2 = 37$, and $R_3 = 46$, as shown in Table 8.6.

TABLE 8.6. AUTHORITARIANISM RANKS OF THREE GROUPS
OF EDUCATORS
(Artificial data)

Teaching-oriented teachers	Administration-oriented teachers	Administrators
4	2	7
9	8	13
3	10	14
1	11	12
5	6	
$R_1 = 22$	$R_2 = 37$	$R_3 = 46$

Now with these data we may compute the value of H from formula (8.1):

$$H = \frac{12}{N(N+1)} \sum_{j=1}^{k} \frac{R_j^2}{n_j} - 3(N+1) \tag{8.1}$$

$$= \frac{12}{14(14+1)} \left[\frac{(22)^2}{5} + \frac{(37)^2}{5} + \frac{(46)^2}{4} \right] - 3(14+1)$$

$$= 6.4$$

Reference to Table O discloses that when the n_j's are 5, 5, and 4, $H \geq 6.4$ has probability of occurrence under the null hypothesis of

$p < .049$. Since this probability is smaller than $\alpha = .05$, our decision in this fictitious study is to reject H_0 in favor of H_1. We conclude that the specified three groups of educators differ in degree of authoritarianism.

Tied observations. When ties occur between two or more scores, each score is given the mean of the ranks for which it is tied.

Since the value of H is somewhat influenced by ties, one may wish to correct for ties in computing H. To correct for the effect of ties, H is computed by formula (8.1) and then divided by

$$1 - \frac{\Sigma T}{N^3 - N} \tag{8.2}$$

where $T = t^3 - t$ (when t is the number of tied observations in a tied group of scores)

N = number of observations in all k samples together, that is, $N = \Sigma n_j$

ΣT directs one to sum over all groups of ties

Thus a general expression for H corrected for ties is

$$H = \frac{\dfrac{12}{N(N+1)} \sum_{j=1}^{k} \dfrac{R_j^2}{n_j} - 3(N+1)}{1 - \dfrac{\Sigma T}{N^3 - N}} \tag{8.3}$$

The effect of correcting for ties is to increase the value of H and thus to make the result more significant than it would have been if uncorrected. Therefore if one is able to reject H_0 without making the correction [i.e., by using formula (8.1) for computing H], one will be able to reject H_0 at an even more stringent level of significance if the correction is used.

In most cases, the effect of the correction is negligible. If no more than 25 per cent of the observations are involved in ties, the probability associated with an H computed without the correction for ties, i.e., by formula (8.1), is rarely changed by more than 10 per cent when the correction for ties is made, that is, if H is computed by formula (8.3), according to Kruskal and Wallis (1952, p. 587).

In the example which follows, H is first computed by formula (8.1) and then is corrected for ties. Notice that even though there are 13 groups of ties, involving 47 of the 56 observations, the change in H which results from applying the correction for ties is merely from $H = 18.464$ to $H = 18.566$.

As usual, the magnitude of the correction factor depends on the length of the ties, i.e., on the values of t, as well as on the percentage of the observations involved.   This point is discussed on page 125.

Example for Large Samples

An investigator determined the birth weights of the members of eight different litters of pigs, in order to determine whether birth weight is affected by litter size.[1]

i. *Null Hypothesis.* H_0: there is no difference in the average birth weights of pigs from litters of different sizes. H_1: the average birth weights of pigs from different litter sizes are not all equal.

ii. *Statistical Test.*   Since the eight litters are independent, a statistical test for k independent samples is appropriate.   Although the measurement of weight in pounds is measurement in a ratio scale, we choose the nonparametric one-way analysis of variance rather than the equivalent parametric test in order to avoid making the assumptions concerning normality and homogeneity of variance associated with the parametric F test and to increase the generality of our findings.

iii. *Significance Level.*   Let $\alpha = .05$. $N = 56 =$ the total number of infant pigs under study.

iv. *Sampling Distribution.*   As computed by formula (8.1), H is distributed approximately as chi square with $df = k - 1$.   Thus the probability associated with the occurrence under H_0 of values as large as an observed H may be determined by reference to Table C.

v. *Rejection Region.*   The region of rejection consists of all values of H which are so large that the probability associated with their occurrence under H_0 for $df = k - 1 = 7$ is equal to or less than $\alpha = .05$.

vi. *Decision.*   The birth weights of the 56 infant pigs belonging to 8 litters are given in Table 8.7.   If we rank these 56 weights, we obtain the ranks shown in Table 8.8.   Observe that we have ranked the 56 scores in a single series, as is required by this test.   The smallest infant pig, the final member of litter 1, weighed 1.1 pounds

---

[1] This example uses an adaptation of the data presented in tables 10.16.1, 10.17.1, 10.20.1, and 10.29.3 of the fifth edition of *Statistical methods* by George W. Snedecor (1956) with the kind permission of the author and the publisher, the Iowa State College Press.   Although these data are not from the behavioral sciences, and although the Kruskal-Wallis test may not be as efficient as a regression analysis in extracting the relevant information in the data, the example is chosen for this illustration because of the large number of ties contained in the observations and because the groups are of unequal size.   This latter feature is rare in research data available in complete form in the current literature of the behavioral sciences.   Kruskal and Wallis (1952) use the same illustrative data in the presentation of their test.

and is given the rank of 1. The heaviest infant pig, also in litter 1, weighed 4.4 pounds; this weight earned the rank of 56. Also shown in Table 8.8 are the sums of each column of ranks, the R_j's.

TABLE 8.7. BIRTH WEIGHTS IN POUNDS OF EIGHT LITTERS OF
POLAND CHINA PIGS, SPRING, 1919

			Litters				
1	2	3	4	5	6	7	8
2.0	3.5	3.3	3.2	2.6	3.1	2.6	2.5
2.8	2.8	3.6	3.3	2.6	2.9	2.2	2.4
3.3	3.2	2.6	3.2	2.9	3.1	2.2	3.0
3.2	3.5	3.1	2.9	2.0	2.5	2.5	1.5 '
4.4	2.3	3.2	3.3	2.0		1.2	
3.6	2.4	3.3	2.5	2.1		1.2	
1.9	2.0	2.9	2.6				
3.3	1.6	3.4	2.8				
2.8		3.2					
1.1		3.2					

With the data in Table 8.8, we may compute the value of H uncorrected for ties, by formula (8.1):

$$H = \frac{12}{N(N+1)} \sum_{j=1}^{k} \frac{R_j^2}{n_j} - 3(N+1) \tag{8.1}$$

$$= \frac{12}{56(56+1)} \left[\frac{(317)^2}{10} + \frac{(216.5)^2}{8} + \frac{(414)^2}{10} + \frac{(277.5)^2}{8} + \frac{(105.5)^2}{6} \right.$$
$$\left. + \frac{(122)^2}{4} + \frac{(71.5)^2}{6} + \frac{(72)^2}{4} \right] - 3(56+1)$$

$$= \frac{12}{3,192} (10{,}048.9 + 5{,}859.031 + 17{,}139.6 + 9{,}625.781$$
$$+ 1{,}855.042 + 3{,}721.0 + 852.042 + 1{,}296.0) - 171$$

$$= 18.464$$

Reference to Table C indicates that an $H \geq 18.464$ with

$$df = k - 1 = 7$$

has probability of occurrence under H_0 of $p < .02$.

To correct for ties, we must first determine how many groups of ties occurred and how many scores were tied in each group. The first tie occurred between two pigs in litter 7 (who both weighed 1.2

pounds). Both were assigned the rank of 2.5. Here $t =$ the number of tied observations $= 2$. For this occurrence,

$$T = t^3 - t = 8 - 2 = 6$$

The next tie occurred between four pigs who were assigned the tied rank of 8.5. Here $t = 4$, and $T = t^3 - t = 64 - 4 = 60$.

TABLE 8.8. RANKS OF BIRTH WEIGHTS OF EIGHT LITTERS OF PIGS

			Litters				
1	2	3	4	5	6	7	8
8.5	52.5	47.5	41.0	23.0	36.0	23.0	18.5
27.5	27.5	54.5	47.5	23.0	31.5	12.5	15.5
47.5	41.0	23.0	41.0	31.5	36.0	12.5	34.0
41.0	52.5	36.0	31.5	8.5	18.5	18.5	4.0
56.0	14.0	41.0	47.5	8.5		2.5	
54.5	15.5	47.5	18.5	11.0		2.5	
6.0	8.5	31.5	23.0				
47.5	5.0	51.0	27.5				
27.5		41.0					
1.0		41.0					
$R_1 = 317.0$	$R_2 = 216.5$	$R_3 = 414.0$	$R_4 = 277.5$	$R_5 = 105.5$	$R_6 = 122.0$	$R_7 = 71.5$	$R_8 = 72.0$

Continuing through the data in Table 8.8 in this way, we find that 13 groups of ties occurred. We may count the number of observations in each tied group to determine the various values of t, and we may compute the value of $T = t^3 - t$ in each case. Our count will result in the findings below:

t	2	4	2	2	4	5	4	4	3	7	6	2	2
T	6	60	6	6	60	120	60	60	24	336	210	6	6

Observe that for any particular value of t, the value of T is a constant. Now, using formula (8.2), we may compute the total correction for ties:

$$1 - \frac{\Sigma T}{N^3 - N} \qquad (8.2)$$

$$= 1 - \frac{(6+60+6+6+60+120+60+60+24+336+210+6+6)}{(56)^3 - 56}$$

$$= .9945$$

Now this value becomes the denominator of formula (8.3), and the value we have already computed from formula (8.1) is the numerator. Thus we need make only one additional operation to compute the value of H corrected for ties:

$$H = \frac{\frac{12}{N(N+1)} \sum_{j=1}^{k} \frac{R_j^2}{n_j} - 3(N+1)}{1 - \frac{\Sigma T}{N^3 - N}} \tag{8.3}$$

$$= \frac{18.464}{.9945}$$

$$= 18.566$$

Reference to Table C discloses that the probability associated with the occurrence under H_0 of a value as large as $H = 18.566$, $df = 7$, is $p < .01$. Since this probability is smaller than our previously set level of significance, $\alpha = .05$, our decision is to reject H_0.* We conclude that the birth weight of pigs varies significantly with litter size.

Summary of procedure. These are the steps in the use of the Kruskal-Wallis one-way analysis of variance by ranks:

1. Rank all of the observations for the k groups in a single series, assigning ranks from 1 to N.

2. Determine the value of R (the sum of the ranks) for each of the k groups of ranks.

3. If a large proportion of the observations are tied, compute the value of H from formula (8.3). Otherwise use formula (8.1).

4. The method for assessing the significance of the observed value of H depends on the size of k and on the size of the groups:

   a. If $k = 3$ and if n_1, n_2, $n_3 \leq 5$, Table O may be used to determine the associated probability under H_0 of an H as large as that observed.

   b. In other cases, the significance of a value as large as the observed value of H may be assessed by reference to Table C, with

$$df = k - 1$$

5. If the probability associated with the observed value of H is equal to or less than the previously set level of significance, α, reject H_0 in favor of H_1.

Power-Efficiency

Compared with the most powerful parametric test, the F test, under conditions where the assumptions associated with the statistical model of

---

* The parametric analysis of variance yields an $F = 2.987$, which for df's of 7 and 48 corresponds with a probability of .011.

the F test are met, the Kruskal-Wallis test has asymptotic efficiency of $\frac{3}{\pi}$ = 95.5 per cent (Andrews, 1954).

The Kruskal-Wallis test is more efficient than the extension of the median test because it utilizes more of the information in the observations, converting the scores into ranks rather than simply dichotomizing them as above and below the median.

References

The reader will find discussions of the one-way analysis of variance by ranks in Kruskal and Wallis (1952) and in Kruskal (1952).

DISCUSSION

Three nonparametric statistical tests for analyzing data from k independent samples were presented in this chapter.

The first of these, the χ^2 test for k independent samples, is useful when the data are in frequencies and when measurement of the variables under study is in a nominal scale or in discrete categories of an ordinal scale. It tests whether the proportions or frequencies in the various categories are independent of the condition (sample) under which they were observed. That is, it tests the null hypothesis that the k samples have come from the same population or from identical populations with respect to the proportion of cases in the various categories.

The second test presented, the extension of the median test, requires at least ordinal measurement of the variable under study. It tests whether k independent samples of scores on that variable could have been drawn from the same population or identical populations with respect to the median.

The Kruskal-Wallis one-way analysis of variance by ranks, the third test discussed, requires at least ordinal measurement of the variable. It tests whether k independent samples could have been drawn from the same continuous population.

We have no choice among these tests if our data are in frequencies rather than scores, i.e., if we have enumeration data, and if the measurement is no stronger than nominal. The χ^2 test for k independent samples is uniquely useful for such data.

The extension of the median test and the Kruskal-Wallis test may both be applied to the same data, i.e., they have similar requirements for the data under test. When the data are such that either test might be used, the Kruskal-Wallis test will be found to be more efficient because it uses more of the information in the observations. It converts the scores to ranks, whereas the extension of the median test converts them simply to either pluses or minuses. Thus the Kruskal-Wallis test preserves the magnitude of the scores more fully than does the extension of the median test. For this reason it is usually more sensitive to differences among the k samples of scores.

The Kruskal-Wallis test seems to be the most efficient of the nonparametric tests for k independent samples. It has power-efficiency of $\frac{3}{\pi} = 95.5$ per cent, when compared with the F test, the most powerful parametric test.

There are at least four other nonparametric tests for k independent samples. These four are rather specialized in their usefulness and therefore have not been presented here. However, the reader might find one of them most valuable in meeting certain specific statistical requirements. The first of these tests, the Whitney extension to the Mann-Whitney test (Whitney, 1951), is a significance test for three samples. It differs from the more general Kruskal-Wallis test in application in that it is designed to test the prediction that the three averages will occur in a specific order. The second of these tests is Mosteller's k-sample test of slippage (Mosteller, 1948; Mosteller and Tukey, 1950). The third is a k-sample runs test (Mood, 1940). Jonckheere (1954) presented the fourth, which is a k-sample test against ordered alternatives, i.e., is designed to test the prediction that the k averages will occur in a specific order.

CHAPTER 9

MEASURES OF CORRELATION AND THEIR
TESTS OF SIGNIFICANCE

In research in the behavioral sciences, we frequently wish to know whether two sets of scores are related, or to know the degree of their relation. Establishing that a correlation exists between two variables may be the ultimate aim of a research, as in some studies of personality dynamics, trait clusters, intragroup similarities, etc. Or establishing a correlation may be but one step in a research having other ends, as is the case when we use measures of correlation to test the reliability of our observations.

This chapter will be devoted to the presentation of nonparametric measures of correlation, and the presentation of statistical tests which determine the probability associated with the occurrence of a correlation as large as the one observed in the sample under the null hypothesis that the variables are unrelated in the population. That is, in addition to presenting measures of association we shall present statistical tests which determine the "significance" of the observed association. The problem of measuring *degree* of association between two sets of scores is quite different in character from that of testing for the *existence* of an association in some population. It is, of course, of some interest to be able to state the degree of association between two sets of scores from a given group of subjects. But it is perhaps of greater interest to be able to say whether or not some observed association in a *sample* of scores indicates that the variables under study are most probably associated in the *population* from which the sample was drawn. The correlation coefficient itself represents the degree of association. Tests of the significance of that coefficient determine, at a stated level of probability, whether the association exists in the population from which a sample was drawn to yield the data from which the coefficient was computed.

In the parametric case, the usual measure of correlation is the Pearson product-moment correlation coefficient r. This statistic requires scores which represent measurement in at least an equal-interval scale. If we wish to test the significance of an observed value of r, we must not only

195

meet the measurement requirement but we must also assume that the scores are from a bivariate normal population.

If, with a given set of data, the measurement requirement of r is not met or the normality assumption is unrealistic, then use may be made of one of the nonparametric correlation coefficients and associated significance tests presented in this chapter. Nonparametric measures of correlation are available for both nominal and ordinal data. The tests make no assumption about the shape of the population from which the scores were drawn. Some assume that the variables have underlying continuity, while others do not even make this assumption. Moreover, the researcher will find that, especially with small samples, the computation of nonparametric measures of correlation and tests of significance is easier than the computation of the Pearson r.

The uses and limitations of each measure will be discussed as the measure is presented. A discussion comparing the merits and uses of the various measures will be offered at the close of the chapter.

THE CONTINGENCY COEFFICIENT: C

Function

The contingency coefficient C is a measure of the extent of association or relation between two sets of attributes. It is uniquely useful when we have only categorical (nominal scale) information about one or both sets of these attributes. That is, it may be used when the information about the attributes consists of an unordered series of frequencies.

To use the contingency coefficient, it is not necessary that we be able to assume underlying continuity for the various categories used to measure either or both sets of attributes. In fact, we do not even need to be able to order the categories in any particular way. The contingency coefficient, as computed from a contingency table, will have the same value regardless of how the categories are arranged in the rows and columns.

Method

To compute the contingency coefficient between scores on two sets of categories, say $A_1, A_2, A_3, \ldots, A_k$, and $B_1, B_2, B_3, \ldots, B_r$, we arrange the frequencies in a contingency table like Table 9.1. The data may consist of any number of categories. That is, one may compute a contingency coefficient from a 2 × 2 table, a 2 × 5 table, a 4 × 4 table, a 3 × 7 table, or any $k \times r$ table.

In such a table, we may enter expected frequencies for each cell (E_{ij}'s) by determining what frequencies would occur if there were no association or correlation between the two variables. The larger is the discrepancy between these expected values and the observed cell values, the larger is the degree of association between the two variables and thus the higher is the value of C.

TABLE 9.1. FORM OF THE CONTINGENCY TABLE FROM WHICH C IS COMPUTED

	A_1	A_2	...	A_k	Total
B_1	(A_1B_1)	(A_2B_1)	...	(A_kB_1)	
B_2	(A_1B_2)	(A_2B_2)	...	(A_kB_2)	
. . .					
B_r	(A_1B_r)	(A_2B_r)	...	(A_kB_r)	
Total					N

The degree of association between two sets of attributes, whether orderable or not, and irrespective of the nature of the variable (it may be either continuous or discrete) or of the underlying distribution of the attribute (the population distribution may be normal or any other shape), may be found from a contingency table of the frequencies by

$$C = \sqrt{\frac{\chi^2}{N + \chi^2}} \tag{9.1}$$

where

$$\chi^2 = \sum_{i=1}^{r} \sum_{j=1}^{k} \frac{(O_{ij} - E_{ij})^2}{E_{ij}} \tag{6.3}$$

and where χ^2 is computed by the method presented earlier (pages 104 to 111).

In other words, in order to compute C, one first computes the value of χ^2 by formula (6.3), and then inserts that value into formula (9.1) to get C.

Example

This computation may be illustrated by reference to data which were first presented in Chap. 8, in the discussion of the χ^2 test for k independent samples. The reader will remember that Hollingshead tested whether the high school curriculums chosen by the youth of Elmtown were dependent on the social class of the youths. Observe that this is a question of the association between frequencies from an unordered series (high school curriculum) and frequencies from an ordered series (social class status). Hollingshead's data are repeated in Table 9.2, a 3 × 4 contingency table. For the data in this table,

TABLE 9.2. FREQUENCY OF ENROLLMENT OF ELMTOWN YOUTHS FROM FIVE SOCIAL CLASSES IN THREE ALTERNATIVE HIGH SCHOOL CURRICULUMS*

Curriculum	Class				Total
	I and II	III	IV	V	
College preparatory	23	40	16	2	81
General	11	75	107	14	207
Commercial	1	31	60	10	102
Total	35	146	183	26	390

* Adapted from Table X of Hollingshead, A. B. 1949. *Elmtown's youth*. New York: Wiley, p. 462, with the kind permission of John Wiley & Sons, Inc.

$\chi^2 = 69.2$. (The computation of χ^2 for these data is given on page 177.) Knowing this, we may determine the value of C by using formula (9.1):

$$C = \sqrt{\frac{\chi^2}{N + \chi^2}} \tag{9.1}$$

$$= \sqrt{\frac{69.2}{390 + 69.2}}$$

$$= .39$$

We determine that the correlation, expressed by a contingency coefficient, between social class position and choice of high school curriculum in Elmtown is $C = .39$.

Testing the Significance of the Contingency Coefficient

The scores or observations with which we deal in research are frequently from individuals in whom we are interested because they constitute a random sample from a population in which we are interested. If we observe a correlation between two sets of attributes in the sample, we may wish to determine whether it is plausible for us to conclude that they are associated in the population which is represented by the sample.

If a group of subjects constitutes a random sample from some population, we may determine whether the association that exists between two sets of scores from that sample indicates that an association exists in the population by testing the association for "significance." In testing the significance of a measure of association, we are testing the null hypothesis that there is no correlation in the population, i.e., that the observed value of the measure of association in the sample could have arisen by chance in a random sample from a population in which the two variables were not correlated.

In order to test the null hypothesis, we usually ascertain the sampling distribution of the statistic (in this case, the measure of association) under H_0. We then use an appropriate statistical test to determine whether our observed value of that statistic can reasonably be thought to have arisen under H_0, referring to some predetermined level of significance. If the probability associated with the occurrence under H_0 of a value as large as our observed value of the statistic is equal to or less than our level of significance, that is, if $p \leq \alpha$, then we decide to reject H_0 and we conclude that the observed association in our sample is not a result of chance but rather represents a genuine relation in the population. However, if the statistical test reveals that it is likely that our observed value might have arisen under H_0, that is, if $p > \alpha$, then we decide that our data do not permit us to conclude that there is a relation between the variables in the population from which the sample was drawn. This method of testing hypotheses should by now be thoroughly familiar to the reader. A fuller discussion of the method is given in Chap. 2, and illustrations of its use occur throughout this book.

Now the reader may know that the Pearson product-moment correlation coefficient r may be tested for significance by exactly the method described above. As he reads further in this chapter, he will discover that various nonparametric measures of association are tested for significance by just such a method. As it happens, however, the contingency coefficient is a special case. One reason that we do not refer to the sampling distribution of C in order to test an observed C for significance is that the mathematical complexities of such a procedure are considerable. A better reason, however, is that in the course of computing the value of C we compute a statistic which itself provides a simple and adequate indication of the significance of C. This statistic, of course, is χ^2. We may test whether an observed C differs significantly from chance simply by determining whether the χ^2 for the data is significant.

For any $k \times r$ contingency table, we may determine the significance of the degree of association (the significance of C) by ascertaining the probability associated with the occurrence under H_0 of values as large as the observed value of χ^2, with $df = (k - 1)(r - 1)$. If that probability is equal to or less than α, the null hypothesis may be rejected at that level of significance. Table C gives the probability associated with the occurrence under H_0 of values as large as an observed χ^2. If the χ^2 for the sample values is significant, then we may conclude that in the population the association between the two sets of attributes is not zero.

Example

We have shown that in Elmtown the relation between adolescents' social class status and their curriculum choice is $C = .39$. In the course of computing C, we determined that $\chi^2 = 69.2$. Now if we consider the adolescents in Hollingshead's group to be a random sample from some population, we may test whether social class status is related to curriculum choice in that population by testing $\chi^2 = 69.2$ for significance. By referring to Table C, we may determine that $\chi^2 \geq 69.2$ with $df = (k - 1)(r - 1) = (4 - 1)(3 - 1) = 6$ has probability of occurrence under H_0 of less than .001. Thus we could reject H_0 at the .001 level of significance, and conclude that social class status and high school curriculum choice are related in the population of which the Elmtown youth are a sample. That is, we conclude that $C = .39$ is significantly different from zero.

Summary of Procedure

These are the steps in the use of the contingency coefficient:

1. Arrange the observed frequencies in a $k \times r$ contingency table, like Table 9.1, where k = the number of categories on which one variable is "scored" and r = the number of categories on which the other variable is "scored."

2. Determine the expected frequency under H_0 for each cell by multiplying the two marginal totals common to that cell and then dividing this product by N, the total number of cases. If more than 20 per cent of the cells have expected frequencies of less than 5, or if any cell has an expected frequency of less than 1, combine categories to increase the expected frequencies which are deficient.

3. Using formula (6.3), compute the value of χ^2 for the data.

4. With this value of χ^2, compute the value of C, using formula (9.1).

5. To test whether the observed value of C indicates that there is an association between the two variables in the population sampled, determine the associated probability under H_0 of a value as large as the observed χ^2 with $df = (k - 1)(r - 1)$ by referring to Table C. If that probability is equal to or less than α, reject H_0 in favor of H_1.

Limitations of the Contingency Coefficient

The wide applicability and relatively easy computation of C may seem to make it an ideal all-round measure of association. This is not the case because of several limitations or deficiencies of the statistic. In general, we may say that it is desirable for correlation coefficients to show at least the following characteristics: (a) where there is a complete lack of any association, the coefficient should vanish, i.e., should equal zero, and (b) when the variables show complete dependence on each other—are perfectly correlated—the coefficient should equal unity, or 1. The contingency coefficient has the first but not the second characteristic: it equals zero when there is no association, but it cannot attain unity. This is the first limitation of C.

The upper limit for the contingency coefficient is a function of the number of categories. When $k = r$, the upper limit for C, that is, the C which would occur for two perfectly correlated variables, is $\sqrt{\dfrac{k-1}{k}}$. For instance, the upper limit of C for a 2×2 table is $\sqrt{\frac{1}{2}} = .707$. For a 3×3 table, the maximum value which C can attain is $\sqrt{\frac{2}{3}} = .816$. The fact that the upper limit of C depends on the sizes of k and r creates the second limitation of C. Two contingency coefficients are not comparable unless they are yielded by contingency tables of the same size.

A third limitation of C is that the data must be amenable to the computation of χ^2 before C may appropriately be used. The reader will remember that the χ^2 test can properly be used only if fewer than 20 per cent of the cells have an expected frequency of less than 5 and no cell has an expected frequency of less than 1.

A fourth limitation of C is that C is not directly comparable to any other measure of correlation, e.g., the Pearson r, the Spearman r_S, or the Kendall τ.

In spite of these limitations, the contingency coefficient is an extremely useful measure of association because of its wide applicability. The contingency coefficient makes no assumptions about the shape of the population of scores, it does not require underlying continuity in the variables under analysis, and it requires only nominal measurement (the least refined variety of measurement) of the variables. Because of this freedom from assumptions and requirements, C may often be used to indicate the degree of relation between two sets of scores to which none of the other measures of association which we have presented is applicable.

Power

Because of its nature and its limitations, we should not expect the contingency coefficient to be very powerful in detecting a relation in the population. However, its ease of computation and its complete freedom from restrictive assumptions recommend its use where other measures of correlation may be inapplicable. Because C is a function of χ^2, its limiting power distribution, like that of χ^2, tends to 1 as N becomes large (Cochran, 1952).

References

For other discussions of the contingency coefficient, the reader is referred to Kendall (1948b, chap. 13) and McNemar (1955, pp. 203–207).

THE SPEARMAN RANK CORRELATION COEFFICIENT: r_s

Function

Of all the statistics based on ranks, the Spearman rank correlation coefficient was the earliest to be developed and is perhaps the best known today. This statistic, sometimes called rho, is here represented by r_s. It is a measure of association which requires that both variables be measured in at least an ordinal scale so that the objects or individuals under study may be ranked in two ordered series.

Rationale

Suppose N individuals are ranked according to two variables. For example, we might arrange a group of students in the order of their scores on the college entrance test and again in the order of their scholastic standing at the end of the freshman year. If the ranking on the entrance test is denoted as $X_1, X_2, X_3, \ldots, X_N$, and the ranking on scholastic standing is represented by $Y_1, Y_2, Y_3, \ldots, Y_N$, we may use a measure of rank correlation to determine the relation between the X's and the Y's.

We can see that the correlation between entrance test ranks and scholastic standing would be perfect if and only if $X_i = Y_i$ for all i's. Therefore, it would seem logical to use the various differences,

$$d_i = X_i - Y_i$$

as an indication of the disparity between the two sets of rankings. Suppose Mary McCord received the top score on the entrance examination but places fifth in her class in scholastic standing. Her d would be -4. John Stanislowski, on the other hand, placed tenth on the entrance examination but leads the class in grades. His d is 9. The magnitude of these various d_i's gives us an idea of how close is the relation between entrance examination scores and scholastic standing. If the relation between the two sets of ranks were perfect, every d would be zero. The larger the d_i's, the less perfect must be the association between the two variables.

Now in computing a correlation coefficient it would be awkward to use the d_i's directly. One difficulty is that the negative d_i's would cancel out the positive ones when we tried to determine the total magnitude of the discrepancy. However, if d_i^2 is employed rather than d_i, this difficulty is circumvented. It is clear that the larger are the various d_i's, the larger will be the value of Σd_i^2.

The derivation of the computing formula for r_S is fairly simple. We shall present it here because it may help to expose the nature of the coefficient, and also because the derivation will reveal other forms by which the formula may be expressed. One of these alternative forms will be used later when we find it necessary to correct the coefficient for the presence of tied scores.

If $x = X - \bar{X}$, where $\bar{X}$ is the mean of the *scores* on the X variable, and if $y = Y - \bar{Y}$, then a general expression for a correlation coefficient is (Kendall, 1948a, chap. 2)

$$r = \frac{\Sigma xy}{\sqrt{\Sigma x^2 \Sigma y^2}} \tag{9.2}$$

in which the sums are over the N values in the sample. Now when the X's and Y's are *ranks*, $r = r_S$, and the sum of the N integers, 1, 2, . . . , N is

$$\Sigma X = \frac{N(N+1)}{2}$$

and the sum of their squares, $1^2, 2^2, \ldots , N^2$ can be shown to be

$$\Sigma X^2 = \frac{N(N+1)(2N+1)}{6}$$

Therefore $\qquad \Sigma x^2 = \Sigma(X - \bar{X})^2 = \Sigma X^2 - \frac{(\Sigma X)^2}{N}$

and $\qquad \Sigma x^2 = \frac{N(N+1)(2N+1)}{6} - \frac{N(N+1)^2}{4} = \frac{N^3 - N}{12} \tag{9.3}$

and similarly $\qquad \Sigma y^2 = \frac{N^3 - N}{12}$

Now $\qquad\qquad d = x - y$
$$d^2 = (x - y)^2 = x^2 - 2xy + y^2$$
$$\Sigma d^2 = \Sigma x^2 + \Sigma y^2 - 2\Sigma xy$$

But formula (9.2) states that

$$r = \frac{\Sigma xy}{\sqrt{\Sigma x^2 \Sigma y^2}} = r_S$$

when the observations are ranked. Therefore

$$\Sigma d^2 = \Sigma x^2 + \Sigma y^2 - 2r_S \sqrt{\Sigma x^2 \Sigma y^2}$$

and thus $\qquad\qquad r_S = \frac{\Sigma x^2 + \Sigma y^2 - \Sigma d^2}{2\sqrt{\Sigma x^2 \Sigma y^2}} \tag{9.4}$

With X and Y in ranks, we may substitute $\Sigma x^2 = \dfrac{N^3 - N}{12} = \Sigma y^2$ into formula (9.4), getting

$$r_S = \frac{\dfrac{N^3 - N}{12} + \dfrac{N^3 - N}{12} - \Sigma d^2}{2\sqrt{\left(\dfrac{N^3 - N}{12}\right)\left(\dfrac{N^3 - N}{12}\right)}} \tag{9.5}$$

$$= \frac{2\left(\dfrac{N^3 - N}{12}\right) - \Sigma d^2}{2\left(\dfrac{N^3 - N}{12}\right)}$$

and

$$r_S = 1 - \frac{\Sigma d^2}{\dfrac{N^3 - N}{6}}$$

$$= 1 - \frac{6\Sigma d^2}{N^3 - N} \tag{9.6}$$

Inasmuch as $d = x - y = (X - \bar{X}) - (Y - \bar{Y}) = X - Y$, since $\bar{X} = \bar{Y}$ in ranks, we may write

$$r_S = 1 - \frac{6\displaystyle\sum_{i=1}^{N} d_i^2}{N^3 - N} \tag{9.7}$$

Formula (9.7) is the most convenient formula for computing the Spearman r_S.

Method

To compute r_S, make a list of the N subjects. Next to each subject's entry, enter his rank for the X variable and his rank for the Y variable. Determine then the various values of $d_i =$ the difference between the two ranks. Square each d_i, and then sum all values of d_i^2 to obtain $\displaystyle\sum_{i=1}^{N} d_i^2$. Then enter this value and the value of N (the number of subjects) directly in formula (9.7).

Example

As part of a study of the effect of group pressures for conformity upon an individual in a situation involving monetary risk, the researchers[1] administered the well-known F scale,[2] a measure of authoritarianism, and a scale designed to measure social status strivings[3] to 12 college students. Information about the correlation between the scores on authoritarianism and those on social status strivings was desired. (Social status strivings were indicated by agreement with such statements as "People shouldn't marry below their social level," "For a date, attending a horse show is better than

TABLE 9.3. SCORES ON AUTHORITARIANISM AND
SOCIAL STATUS STRIVINGS

Student	Score	
	Authoritarianism	Social status strivings
A	82	42
B	98	46
C	87	39
D	40	37
E	116	65
F	113	88
G	111	86
H	83	56
I	85	62
J	126	92
K	106	54
L	117	81

going to a baseball game," and "It is worthwhile to trace back your family tree.") Table 9.3 gives each of the 12 students' scores on the two scales.

In order to compute the Spearman rank correlation between these two sets of scores, it was necessary to rank them in two series. The ranks of the scores given in Table 9.3 are shown in Table 9.4, which also shows the various values of d_i and $d_i{}^2$. Thus, for example, Table 9.4 shows that the student (student J) who showed the most authoritarianism (on the F scale) also showed the most extreme social status strivings, and thus was assigned a rank of 12 on both variables. The reader will observe that no student's rank on one

---

[1] Siegel, S., and Fagan, Joen. The Asch effect under conditions of risk. Unpublished study. The data reported here are from a pilot study.

[2] Presented in Adorno, T. W., Frenkel-Brunswik, Else, Levinson, D. J., and Sanford, R. N. 1950. *The authoritarian personality.* New York: Harper.

[3] Siegel, Alberta E., and Siegel, S. An experimental test of some hypotheses in reference group theory. Unpublished study.

TABLE 9.4. RANKS ON AUTHORITARIANISM AND SOCIAL STATUS STRIVINGS

Student	Rank		d_i	d_i^2
	Authoritarianism	Social status strivings		
A	2	3	−1	1
B	6	4	2	4
C	5	2	3	9
D	1	1	0	0
E	10	8	2	4
F	9	11	−2	4
G	8	10	−2	4
H	3	6	−3	9
I	4	7	−3	9
J	12	12	0	0
K	7	5	2	4
L	11	9	2	4
				$\Sigma d_i^2 = 52$

variable was more than three ranks distant from his rank on the other variable, i.e., the largest d_i is 3.

From the data shown in Table 9.4, we may compute the value of r_S by applying formula (9.7):

$$r_S = 1 - \frac{6 \sum_{i=1}^{N} d_i^2}{N^3 - N} \qquad (9.7)$$

$$= 1 - \frac{6(52)}{(12)^3 - 12}$$

$$= .82$$

We observe that for these 12 students the correlation between authoritarianism and social status strivings is $r_S = .82$.

Tied observations. Occasionally two or more subjects will receive the same score on the same variable. When tied scores occur, each of them is assigned the average of the ranks which would have been assigned had no ties occurred, our usual procedure for assigning ranks to tied observations.

If the proportion of ties is not large, their effect on r_S is negligible, and formula (9.7) may still be used for computation. However, if the proportion of ties is large, then a correction factor must be incorporated in the computation of r_S.

The effect of tied ranks in the X variable is to reduce the sum of squares, Σx^2, below the value of $\dfrac{N^3 - N}{12}$, that is,

$$\Sigma x^2 < \frac{N^3 - N}{12}$$

when there are tied ranks in the X variable. Therefore it is necessary to correct the sum of squares, taking ties into account. The correction factor is T:

$$T = \frac{t^3 - t}{12}$$

where t = the number of observations tied at a given rank. When the sum of squares is corrected for ties, it becomes

$$\Sigma x^2 = \frac{N^3 - N}{12} - \Sigma T$$

where ΣT indicates that we sum the various values of T for all the various groups of tied observations.

When a considerable number of ties are present, one uses formula (9.4) (page 203) in computing r_S:

$$r_S = \frac{\Sigma x^2 + \Sigma y^2 - \Sigma d^2}{2\sqrt{\Sigma x^2 \Sigma y^2}} \tag{9.4}$$

where $\Sigma x^2 = \dfrac{N^3 - N}{12} - \Sigma T_x$

$\Sigma y^2 = \dfrac{N^3 - N}{12} - \Sigma T_y$

Example with Ties

In the study cited in the previous example, each student was individually observed in the well-known group pressures situation developed by Asch.[1] In this situation, a group of subjects are asked individually to state which of a group of alternative lines is the same length as a standard line. All but one of these subjects are confederates of the experimenter, and on certain trials they unanimously choose an incorrect match. The naïve subject, who is so seated that he is the last asked to report his judgment, has the choice of standing alone in selecting the true match (which is unmistakable to people in situations where no contradictory group pressures are involved) or of "yielding" to group pressures by stating that the incorrect line is the match.

The modification which Siegel and Fagan introduced into this

[1] Asch, S. E. 1952. *Social psychology.* New York: Prentice-Hall, pp. 451–476.

experiment was to agree to pay each subject 50 cents for every correct judgment, and to penalize him 50 cents for every incorrect one. The subjects were given $2 at the start of the experiment, and they understood that they could keep all moneys in their possession at the end of the session. So far as the naïve subject knew, this agreement held with all members of the group making the judgments. Each naïve subject participated in 12 "crucial" trials, i.e., in 12 trials in which the confederates unanimously chose the wrong line as the match. Thus each naïve subject could "yield" up to 12 times.

TABLE 9.5. SCORES ON YIELDING AND SOCIAL STATUS STRIVINGS

Student	Number of yieldings	Social status strivings score
A	0	42
B	0	46
C	1	39
D	1	37
E	3	65
F	4	88
G	5	86
H	6	56
I	7	62
J	8	92
K	8	54
L	12	81

As part of the study, the experimenters wanted to know whether yielding in this situation is correlated with social status strivings, as measured by the scale described previously. This was determined by computing a Spearman rank correlation between the scores of each of the 12 naïve subjects on the social status strivings scale and the number of times that each yielded to the group pressures. The data on these two variables are presented in Table 9.5. Observe that two of the naïve subjects did not yield at all (these were students A and B), whereas one (student L) yielded on every crucial trial. The scores presented in Table 9.5 are ranked in Table 9.6· Observe that there are 3 sets of tied observations on the X variable (number of yieldings). Two subjects tied at 0; both were given ranks of 1.5. Two tied at 1; both were given ranks of 3.5. And two tied at 8: both were given ranks of 10.5.

Because of the relatively large proportion of tied observations in the X variable, it might be felt that formula (9.4) should be used in

TABLE 9.6. RANKS FOR YIELDING AND SOCIAL STATUS STRIVINGS

| Student | Rank | | d_i | $d_i{}^2$ |
	Yielding	Social status strivings		
A	1.5	3	−1.5	2.25
B	1.5	4	−2.5	6.25
C	3.5	2	1.5	2.25
D	3.5	1	2.5	6.25
E	5	8	−3.0	9.00
F	6	11	−5.0	25.00
G	7	10	−3.0	9.00
H	8	6	2.0	4.00
I	9	7	2.0	4.00
J	10.5	12	−1.5	2.25
K	10.5	5	−5.5	30.25
L	12	9	3.0	9.00
				$\Sigma d_i{}^2 = 109.50$

computing the value of r_S. To use that formula, we must first determine the values of Σx^2 and Σy^2.

Now with 3 sets of tied observations on the X variable, where $t = 2$ in each set, we have

$$\Sigma x^2 = \frac{N^3 - N}{12} - \Sigma T_x$$
$$= \frac{(12)^3 - 12}{12} - \left(\frac{2^3 - 2}{12} + \frac{2^3 - 2}{12} + \frac{2^3 - 2}{12} \right)$$
$$= 143 - 1.5$$
$$= 141.5$$

That is, corrected for ties, $\Sigma x^2 = 141.5$. We find Σy^2 by a comparable method:

$$\Sigma y^2 = \frac{N^3 - N}{12} - \Sigma T_y$$

But inasmuch as there are no ties in the Y scores (the scores on social status strivings), $\Sigma T_y = 0$, and thus

$$\Sigma y^2 = \frac{(12)^3 - 12}{12} - 0$$
$$= 143$$

Corrected for ties, $\Sigma x^2 = 141.5$ and $\Sigma y^2 = 143$. From the addition shown in Table 9.6, we know that $\Sigma d_i^2 = 109.5$. Substituting these values in formula (9.4), we have

$$r_S = \frac{\Sigma x^2 + \Sigma y^2 - \Sigma d^2}{2\sqrt{\Sigma x^2 \Sigma y^2}} \tag{9.4}$$

$$= \frac{141.5 + 143 - 109.5}{2\sqrt{(141.5)(143)}}$$

$$= .616$$

Corrected for ties, the correlation between amount of yielding and degree of social status strivings is $r_S = .616$. If we had computed r_S from formula (9.7), i.e., if we had not corrected for ties, we would have found $r_S = .617$. This illustrates the relatively insignificant effect of ties upon the value of the Spearman rank correlation. Notice, however, that the effect of ties is to inflate the value of r_S. For this reason, the correction should be used where there is a large proportion of ties in either or both the X and Y variables.

Testing the Significance of r_S

If the subjects whose scores were used in computing r_S were randomly drawn from some population, we may use those scores to determine whether the two variables are associated in the population. That is, we may wish to test the null hypothesis that the two variables under study are *not* associated in the population and that the observed value of r_S differs from zero only by chance.

Small samples. Suppose that the null hypothesis is true. That is, suppose that there is no relation in the population between the X and Y variables. Now if a sample of X and Y scores is randomly drawn from that population, for a given rank order of the Y scores any rank order of the X scores is just as likely as any other rank order of the X scores. And for any given order of the X scores, all possible orders of the Y scores are equally likely. For N subjects, there are $N!$ possible rankings of X scores which may occur in association with any given ranking of Y scores. Since these are equally likely, the probability of the occurrence of any particular ranking of the X scores with a given ranking of the Y scores is $\frac{1}{N!}$.

For each of the possible rankings of Y there will be associated a value of r_S. The probability of the occurrence under H_0 of any particular value of r_S is thus proportional to the number of permutations giving rise to that value.

Using formula (9.7), the computation formula for r_S, we find that for

$N = 2$, only two values of r_S are possible: $+1$ and -1. Each of these has probability of occurrence under H_0 of $\frac{1}{2}$.

For $N = 3$, the possible values of r_S are -1, $-\frac{1}{2}$, $+\frac{1}{2}$, and $+1$. Their respective probabilities under H_0 are $\frac{1}{6}$, $\frac{1}{3}$, $\frac{1}{3}$, and $\frac{1}{6}$.

Table P of the Appendix gives critical values of r_S which have been arrived at by a similar method. For N from 4 to 30, the table gives the value of r_S which has an associated probability under H_0 of $p = .05$ and the value of r_S which has an associated probability under H_0 of $p = .01$. This is a one-tailed table, i.e., the stated probabilities apply when the observed value of r_S is in the predicted direction, either positive or negative. If an observed value of r_S equals or exceeds the value tabled, that observed value is significant (for a one-tailed test) at the level indicated.

Example[1]

We have already found that for $N = 12$ the correlation between authoritarianism and social status strivings is $r_S = .82$. Table P shows that a value as large as this is significant at the $p < .01$ level (one-tailed test). Thus we could reject H_0 at the $\alpha = .01$ level, concluding that in the population of students from which the sample was drawn, authoritarianism and social status strivings are associated.

---

[1] In testing a measure of association for significance, we follow the same six steps which we have followed in all other statistical tests throughout this book. That is, (i) the null hypothesis is that the two variables are unrelated in the population, whereas H_1 is that they are related or associated in the population; (ii) the statistical test is the significance test appropriate for the measure of association; (iii) the level of significance is specified in advance, and may be any small probability, for example, $\alpha = .05$ or $\alpha = .01$, etc., while the N is the number of cases which have yielded scores on both variables; (iv) the sampling distribution is the theoretical distribution of the measure under H_0, exact probabilities or critical values from which are given in the tables used to test the measure for significance; (v) the rejection region consists of all values of the measure of association which are so extreme that the probability associated with their occurrence under H_0 is equal to or less than α (and a one-tailed region of rejection is used when the *sign* of the association is predicted in H_1), and (vi) the decision consists of determining the observed value of the measure of association and then determining the probability under H_0 associated with such an extreme value; if and only if that probability is equal to or less than α, the decision is to reject H_0 in favor of H_1.

Because the same sets of data are repeatedly used for illustrative material in the discussions of the various measures of association, in order to highlight the differences and similarities among these measures, the constant repetition of the six steps of statistical inference in the examples would lead to unnecessary redundancy. Therefore we have chosen not to include these six steps in the presentation of the examples in this chapter. We mention here that they might well have been included in order to point out to the reader that the decision-making procedure used in testing the significance of a measure of association is identical to the decision-making procedure used in other sorts of statistical tests.

We have also seen that the relation between social status strivings and amount of yielding is $r_S = .62$ in our group of 12 subjects. By referring to Table P, we can determine that $r_S \geq .62$ has probability of occurrence under H_0 between $p = .05$ and $p = .01$ (one-tailed). Thus we could conclude, at the $\alpha = .05$ level, that these two variables are associated in the population from which the sample was drawn.

Large samples. When N is 10 or larger, the significance of an obtained r_S under the null hypothesis may be tested by (Kendall, 1948a, pp. 47–48):

$$t = r_S \sqrt{\frac{N - 2}{1 - r_S^2}} \qquad (9.8)$$

That is, for N large, the value defined by formula (9.8) is distributed as Student's t with $df = N - 2$. Thus the associated probability under H_0 of any value as extreme as an observed value of r_S may be determined by computing the t associated with that value, using formula (9.8), and then determining the significance of that t by referring to Table B of the Appendix.

Example[1]

We have already determined that the relation between social status strivings and amount of yielding is $r_S = .62$ for $N = 12$. Since this N is larger than 10, we may use the large sample method of testing this r_S for significance:

$$t = .62 \sqrt{\frac{12 - 2}{1 - (.62)^2}}$$

$$= 2.49$$

Table B shows that for $df = N - 2 = 12 - 2 = 10$, a t as large as 2.48 is significant at the .025 level but not at the .01 level for a one-tailed test. This is essentially the same result we obtained previously by using Table P. We could reject H_0 at $\alpha = .05$, concluding that social status strivings and amount of yielding are associated in the population of which the 12 students were a sample.

Summary of Procedure

These are the steps in the use of the Spearman rank correlation coefficient:

1. Rank the observations on the X variable from 1 to N. Rank the observations on the Y variable from 1 to N.

[1] See footnote, page 211.

2. List the N subjects. Give each subject's rank on the X variable and his rank on the Y variable next to his entry.

3. Determine the value of d_i for each subject by subtracting his Y rank from his X rank. Square this value to determine each subject's d_i^2. Sum the d_i^2's for the N cases to determine Σd_i^2.

4. If the proportion of ties in either the X or the Y observations is large, use formula (9.4) to compute r_S. In other cases, use formula (9.7).

5. If the subjects constitute a random sample from some population, one may test whether the observed value of r_S indicates an association between the X and Y variables in the population. The method for doing so depends on the size of N:

    a. For N from 4 to 30, critical values of r_S for the .05 and .01 levels of significance (one-tailed test) are shown in Table P.

    b. For $N \geq 10$, the significance of a value as large as the observed value of r_S may be determined by computing the t associated with that value [using formula (9.8)] and then determining the significance of that value of t by referring to Table B.

Power-Efficiency

The efficiency of the Spearman rank correlation when compared with the most powerful parametric correlation, the Pearson r, is about 91 per cent (Hotelling and Pabst, 1936). That is, when r_S is used with a sample to test for the existence of an association in the population, and when the assumptions and requirements underlying the proper use of the Pearson r are met, that is, when the population has a bivariate normal distribution and measurement is in the sense of at least an interval scale, then r_S is 91 per cent as efficient as r in rejecting H_0. If a correlation between X and Y exists in that population, with 100 cases r_S will reveal that correlation at the same level of significance which r attains with 91 cases.

References

For other discussions of the Spearman rank-order correlation, the reader may turn to Hotelling and Pabst (1936), Kendall (1948a; 1948b, chap. 16), and Olds (1949).

THE KENDALL RANK CORRELATION COEFFICIENT: τ

Function

The Kendall rank correlation coefficient, τ (tau), is suitable as a measure of correlation with the same sort of data for which r_S is useful. That is, if at least ordinal measurement of both the X and Y variables has been achieved, so that every subject can be assigned a rank on both X and Y, then τ will give a measure of the degree of association or correlation between the two sets of ranks. The sampling distribution of τ under the null hypothesis is known, and therefore τ, like r_S, is subject to tests of significance.

One advantage of τ over r_S is that τ can be generalized to a partial correlation coefficient. This partial coefficient will be presented in the section following this one.

Rationale

Suppose we ask judge X and judge Y to rank four objects. For example, we might ask them to rank four essays in order of quality of expository style. We represent the four papers as a, b, c, and d. The obtained rankings are these:

Essay	a	b	c	d
Judge X	3	4	2	1
Judge Y	3	1	4	2

If we rearrange the order of the essays so that judge X's ranks appear in natural order (i.e., 1, 2, . . . , N), we get

Essay	d	c	a	b
Judge X	1	2	3	4
Judge Y	2	4	3	1

We are now in a position to determine the degree of correspondence between the judgments of X and Y. Judge X's rankings being in their natural order, we proceed to determine how many pairs of ranks in judge Y's set are in their correct (natural) order with respect to each other.

Consider first all possible *pairs* of ranks in which judge Y's rank 2, the rank farthest to the left in his set, is one member. The first pair, 2 and 4, has the correct order: 2 precedes 4. Since the order is "natural," we assign a score of $+1$ to this pair. Ranks 2 and 3 constitute the second pair. This pair is also in the correct order, so it also earns a score of $+1$. Now the third pair consists of ranks 2 and 1. These ranks are not in "natural" order; 2 precedes 1. Therefore we assign this pair a score of -1. For all pairs which include the rank 2, we total the scores:

$$(+1) + (+1) + (-1) = +1$$

Now we consider all possible pairs of ranks which include rank 4 (which is the rank second from the left in judge Y's set) and one succeeding rank. One pair is 4 and 3; the two members of the pair are not in the natural order, so the score for that pair is -1. Another pair is 4 and 1; again a score of -1 is assigned. The total of these scores is

$$(-1) + (-1) = -2$$

When we consider rank 3 and succeeding ranks, we get only this pair: 3 and 1. The two members of this pair are in the wrong order; therefore this pair receives a score of -1.

The total of all the scores we have assigned is

$$(+1) + (-2) + (-1) = -2$$

Now what is the *maximum possible* total we could have obtained for the scores assigned all the pairs in judge Y's ranking? The maximum possible total would have been yielded if the rankings of judges X and Y had agreed perfectly, for then, when the rankings of judge X were arranged in their natural order, every pair of judge Y's ranks would also be in the correct order and thus every pair would receive a score of $+1$. The maximum possible total then, the one which would occur in the case of perfect agreement between X and Y, would be four things taken two at a time, or $\binom{4}{2} = 6$.

The degree of relation between the two sets of ranks is indicated by the ratio of the actual total of $+1$'s and -1's to the possible maximum total. The Kendall rank correlation coefficient is that ratio:

$$\tau = \frac{\text{actual total}}{\text{maximum possible total}} = \frac{-2}{6} = -.33$$

That is, $\tau = -.33$ is a measure of the agreement between the ranks assigned to the essays by judge X and those assigned by judge Y.

One may think of τ as a function of the minimum number of inversions or interchanges between neighbors which is required to transform one ranking into another. That is, τ is a sort of coefficient of disarray.

Method

We have seen that

$$\tau = \frac{\text{actual score}}{\text{maximum possible score}}$$

In general, the maximum possible score will be $\binom{N}{2}$, which can be expressed as $\frac{1}{2}N(N-1)$. Thus this last expression may be the denominator of the formula for τ. For the numerator, let us denote the observed sum of the $+1$ and -1 scores for all pairs as S. Then

$$\tau = \frac{S}{\frac{1}{2}N(N-1)} \tag{9.9}$$

where $N = $ the number of objects or individuals ranked on both X and Y.

The calculation of S may be shortened considerably from the method shown above in the discussion of the logic of the measure.

When the ranks of judge X were in the natural order, the corresponding ranks of judge Y were in this order

$$\text{Judge } Y: \quad 2 \quad 4 \quad 3 \quad 1$$

We can determine the value of S by starting with the first number on the left and counting the number of ranks to its right which are larger. We then subtract from this the number of ranks to its right which are smaller. If we do this for all ranks and then sum the results, we obtain S.

Thus, for the above set of ranks, to the right of rank 2 are ranks 3 and 4 which are larger and rank 1 which is smaller. Rank 2 thus contributes $(+2-1) = +1$ to S. For rank 4, no ranks to its right are larger but two (ranks 3 and 1) are smaller. Rank 4 thus contributes $(0-2) = -2$ to S. For rank 3, no rank to its right is larger but one (rank 1) is smaller, so rank 3 contributes $(0-1) = -1$ to S. These contributions total

$$(+1) + (-2) + (-1) = -2 = S$$

Knowing S, we may use formula (9.9) to compute the value of τ for the ranks assigned by the two judges:

$$\tau = \frac{S}{\frac{1}{2}N(N-1)} \tag{9.9}$$

$$= \frac{-2}{\frac{1}{2}(4)(4-1)}$$

$$= -.33$$

Example

We have already computed the Spearman r_S for 12 students' scores on authoritarianism and on social status strivings. The scores of the 12 students are presented in Table 9.3, and the ranks of these scores are presented in Table 9.4. We may compute the value of τ for the same data.

The two sets of ranks to be correlated (shown in Table 9.4) are these:

Subject	A	B	C	D	E	F	G	H	I	J	K	L
Status strivings rank	3	4	2	1	8	11	10	6	7	12	5	9
Authoritarianism rank	2	6	5	1	10	9	8	3	4	12	7	11

To compute τ, we shall rearrange the order of the subjects so that the rankings on social status strivings occur in the natural order:

Subject	D	C	A	B	K	H	I	E	L	G	F	J
Status strivings rank	1	2	3	4	5	6	7	8	9	10	11	12
Authoritarianism rank	1	5	2	6	7	3	4	10	11	8	9	12

Having arranged the ranks on variable X in their natural order, we determine the value of S for the corresponding order of ranks on variable Y:

$$S = (11 - 0) + (7 - 3) + (9 - 0) + (6 - 2) + (5 - 2)$$
$$+ (6 - 0) + (5 - 0) + (2 - 2) + (1 - 2) + (2 - 0)$$
$$+ (1 - 0) = 44$$

The authoritarianism rank which is farthest to the left is 1. This rank has 11 ranks which are larger to its right, and 0 ranks which are smaller, so its contribution to S is $(11 - 0)$. The next rank is 5. It has 7 ranks to its right which are larger and 3 to its right which are smaller, so that its contribution to S is $(7 - 3)$. By proceeding in this way, we obtain the various values shown above, which we have summed to yield $S = 44$. Knowing that $S = 44$ and $N = 12$, we may use formula (9.9) to compute τ:

$$\tau = \frac{S}{\frac{1}{2}N(N - 1)} \qquad (9.9)$$

$$= \frac{44}{\frac{1}{2}(12)(12 - 1)}$$

$$= .67$$

$\tau = .67$ represents the degree of relation between authoritarianism and social status strivings shown by the 12 students.

Tied observations. When two or more observations on either the X or the Y variable are tied, we turn to our usual procedure in ranking tied scores: the tied observations are given the average of the ranks they would have received if there were no ties.

The effect of ties is to change the denominator of our formula for τ. In the case of ties, τ becomes

$$\tau = \frac{S}{\sqrt{\frac{1}{2}N(N-1) - T_X}\ \sqrt{\frac{1}{2}N(N-1) - T_Y}} \tag{9.10}$$

where $T_X = \frac{1}{2}\Sigma t(t-1)$, t being the number of tied observations in each group of ties on the X variable

$T_Y = \frac{1}{2}\Sigma t(t-1)$, t being the number of tied observations in each group of ties on the Y variable

The computations required by formula (9.10) are illustrated in the example which follows.

Example with Ties

Again we shall repeat an example which was first presented in the discussion of the Spearman r_S. We correlated the scores of 12 subjects on a scale measuring social status strivings with the number of times that each yielded to group pressures in judging the length of lines. The data for this pilot study are presented in Table 9.5. These scores are converted to ranks in Table 9.6.

The two sets of ranks to be correlated (first presented in Table 9.6) are these:

Subject	A	B	C	D	E	F	G	H	I	J	K	L
Status strivings rank	3	4	2	1	8	11	10	6	7	12	5	9
Yielding rank	1.5	1.5	3.5	3.5	5	6	7	8	9	10.5	10.5	12

As usual, we first rearrange the order of the subjects, so that the ranks on the X variable occur in natural order:

Subject	D	C	A	B	K	H	I	E	L	G	F	J
Status strivings rank	1	2	3	4	5	6	7	8	9	10	11	12
Yielding rank	3.5	3.5	1.5	1.5	10.5	8	9	5	12	7	6	10.5

Then we compute the value of S in the usual way:

$$S = (8 - 2) + (8 - 2) + (8 - 0) + (8 - 0) + (1 - 5)$$
$$+ (3 - 3) + (2 - 3) + (4 - 0) + (0 - 3) + (1 - 1)$$
$$+ (1 - 0) = 25$$

Having determined that $S = 25$, we now determine the values of T_X and T_Y. There are no ties among the scores on social status strivings, i.e., in the X ranks, and thus $T_X = 0$.

On the Y variable (yielding), there are three sets of tied ranks. Two subjects are tied at rank 1.5, two are tied at 3.5, and two are tied at 10.5. In each of these cases, $t = 2$, the number of tied observations. Thus T_Y may be computed:

$$T_Y = \tfrac{1}{2}\Sigma t(t - 1)$$
$$= \tfrac{1}{2}[2(2 - 1) + 2(2 - 1) + 2(2 - 1)]$$
$$= 3$$

With $T_X = 0$, $T_Y = 3$, $S = 25$, and $N = 12$, we may determine the value of τ by using formula (9.10):

$$\tau = \frac{S}{\sqrt{\tfrac{1}{2}N(N - 1) - T_X}\,\sqrt{\tfrac{1}{2}N(N - 1) - T_Y}} \qquad (9.10)$$

$$= \frac{25}{\sqrt{\tfrac{1}{2}(12)(12 - 1) - 0}\,\sqrt{\tfrac{1}{2}(12)(12 - 1) - 3}}$$

$$= .39$$

If we had not corrected the above coefficient for ties, i.e., if we had used formula (9.9) in computing τ, we would have found $\tau = .38$. Observe that the effect of correcting for ties is relatively small.

Comparison of τ and r_S

In two cases we have computed both τ and r_S for the same data. The reader will have noted that the numerical values of τ and r_S are not identical when both are computed from the same pair of rankings. For the relation between authoritarianism and social status strivings, $r_S = .82$ whereas $\tau = .67$. For the relation between social status strivings and number of yieldings to group pressures, $r_S = .62$ and $\tau = .39$.

These examples illustrate the fact that τ and r_S have different underlying scales, and numerically they are not directly comparable to each other. That is, if we measure the degree of correlation between the variables A and B by using r_S, and then do the same for A and C by using τ, we cannot then say whether A is more closely related to B or to C, for we shall be using two noncomparable measures of correlation.

However, both coefficients utilize the same amount of information in the data, and thus both have the same power to detect the existence of association in the population. That is, the sampling distributions of τ and r_S are such that with a given set of data both will reject the null hypothesis (that the variables are unrelated in the population) at the same level of significance. This should become clearer after the following discussion on testing the significance of τ.

Testing the Significance of τ

If a random sample is drawn from some population in which X and Y are unrelated, and the members of the sample are ranked on X and Y, then for any given order of the X ranks all possible orders of the Y ranks are equally likely. That is, for a given order of the X ranks, any one possible order of the Y ranks is just as likely to occur as any other possible order of the Y ranks. Suppose we order the X ranks in natural order, i.e., 1, 2, 3, . . . , N. For that order of the X ranks, all the $N!$ possible orders of the Y ranks are equally probable under H_0. Therefore any particular order of the Y ranks has probability of occurrence under H_0 of $1/N!$.

TABLE 9.7. PROBABILITIES OF τ UNDER H_0 FOR $N = 4$

Value of τ	Frequency of occurrence under H_0	Probability of occurrence under H_0
-1.0	1	$\frac{1}{24}$
$-.67$	3	$\frac{3}{24}$
$-.33$	5	$\frac{5}{24}$
0	6	$\frac{6}{24}$
$.33$	5	$\frac{5}{24}$
$.67$	3	$\frac{3}{24}$
1.0	1	$\frac{1}{24}$

For each of the $N!$ possible rankings of Y, there will be associated a value of τ. These possible values of τ will range from $+1$ to -1, and they can be cast in a frequency distribution. For instance, for $N = 4$ there are $4! = 24$ possible arrangements of the Y ranks, and each has an associated value of τ. Their frequency of occurrence under H_0 is shown in Table 9.7.

We could compute similar tables of probabilities for other values of N, but of course as N increases this method becomes increasingly tedious.

Fortunately, for $N \geq 8$, the sampling distribution of τ is practically indistinguishable from the normal distribution (Kendall, 1948a, pp. 38–39). Therefore, for N large, we may use the normal curve table (Table A) for determining the probability associated with the occurrence under H_0 of any value as extreme as an observed value of τ.

However, when N is 10 or less, Table Q of the Appendix may be used to determine the exact probability associated with the occurrence (one-tailed) under H_0 of any value as extreme as an observed S. (The sampling distributions of S and τ are identical, in a probability sense.

Inasmuch as τ is a function of S, either might be tabled.   It is more convenient to tabulate S.)   For such small samples, the significance of an observed relation between two samples of ranks may be determined by simply finding the value of S and then referring to Table Q to determine the probability (one-tailed) associated with that value.   If the $p \le \alpha$, H_0 may be rejected.   For example, suppose $N = 8$ and $S = 10$.   Table Q shows that an $S \ge 10$ for $N = 8$ has probability of occurrence under H_0 of $p = .138$.

When N is larger than 10, τ may be considered to be normally distributed with

$$\text{Mean} = \mu_\tau = 0$$

and $\qquad$ Standard deviation $= \sigma_\tau = \sqrt{\dfrac{2(2N + 5)}{9N(N - 1)}}$

That is,

$$z = \frac{\tau - \mu_\tau}{\sigma_\tau} = \frac{\tau}{\sqrt{\dfrac{2(2N + 5)}{9N(N - 1)}}} \tag{9.11}$$

is approximately normally distributed with zero mean and unit variance. Thus the probability associated with the occurrence under H_0 of any value as extreme as an observed τ may be determined by computing the value of z as defined by formula (9.11) and then determining the significance of that z by reference to Table A of the Appendix.

Example for $N > 10$*

We have already determined that among 12 students the correlation between authoritarianism and social status strivings is $\tau = .67$. If we consider these 12 students to be a random sample from some population, we may test whether these two variables are associated in that population by using formula (9.11):

$$z = \frac{\tau}{\sqrt{\dfrac{2(2N + 5)}{9N(N - 1)}}} \tag{9.11}$$

$$= \frac{.67}{\sqrt{\dfrac{2[(2)(12) + 5]}{(9)(12)(12 - 1)}}}$$

$$= 3.03$$

By referring to Table A, we see that $z \ge 3.03$ has probability of occurrence under H_0 of $p = .0012$.   Thus we could reject H_0 at

* See footnote, page 211.

level of significance α = .0012, and conclude that the two variables are associated in the population from which this sample was drawn.

We have already mentioned that τ and r_S have identical power to reject H_0. That is, even though τ and r_S are numerically different for the same set of data, their sampling distributions are such that with the same data H_0 would be rejected at the same level of significance by the significance tests associated with both measures.

In the present case, τ = .67. Associated with this value is z = 3.03, which permits us to reject H_0 at α = .0012. When the Spearman coefficient was computed from the same data, we found r_S = .82. When we apply to that value the significance test for r_S [formula (9.8)], we arrive at t = 4.53 with df = 10. Table B shows that $t \geq 4.53$ with df = 10 has probability of occurrence under H_0 of slightly higher than .001. Thus τ and r_S for the same set of data have significance tests which reject H_0 at essentially the same level of significance.

Summary of Procedure

These are the steps in the use of the Kendall rank correlation coefficient:

1. Rank the observations on the X variable from 1 to N. Rank the observations on the Y variable from 1 to N.

2. Arrange the list of N subjects so that the X ranks of the subjects are in their natural order, that is, 1, 2, 3, . . . , N.

3. Observe the Y ranks in the order in which they occur when the X ranks are in natural order. Determine the value of S for this order of the Y ranks.

4. If there are no ties among either the X or the Y observations, use formula (9.9) in computing the value of τ. If there are ties, use formula (9.10).

5. If the N subjects constitute a random sample from some population, one may test whether the observed value of τ indicates the existence of an association between the X and Y variables in that population. The method for doing so depends on the size of N:

    a. For $N \leq 10$, Table Q shows the associated probability (one-tailed) of a value as large as an observed S.

    b. For $N > 10$, one may compute the value of z associated with τ by using formula (9.11). Table A shows the associated probability of a value as large as an observed z.

If the p yielded by the appropriate method is equal to or less than α, H_0 may be rejected in favor of H_1.

Power-Efficiency

The Spearman r_S and the Kendall τ are equally powerful in rejecting H_0, inasmuch as they make equivalent use of the information in the data.

When used on data to which the Pearson r is properly applicable, both τ and r_S have efficiency of 91 per cent. That is, τ is approximately as sensitive a test of the existence of association between two variables in a bivariate normal population with a sample of 100 cases as is the Pearson r with 91 cases (Hotelling and Pabst, 1936; Moran, 1951).

References

The reader will find other discussions of the Kendall τ in Kendall (1938; 1945; 1947; 1948a; 1948b; 1949).

THE KENDALL PARTIAL RANK CORRELATION COEFFICIENT: $\tau_{xy.z}$

Function

When correlation is observed between two variables, there is always the possibility that this correlation is due to the association between each of the two variables and a third variable. For example, among a group of school children of diverse ages, one might find a high correlation between size of vocabulary and height. This correlation may not reflect any genuine or direct relation between these two variables, but rather may result from the fact that both vocabulary size and height are associated with a third variable, age.

Statistically, this problem may be attacked by methods of partial correlation. In partial correlation, the effects of variation by a third variable upon the relation between the X and Y variables are eliminated. In other words, the correlation between X and Y is found with the third variable Z kept constant.

In designing an experiment, one has the alternative of either introducing experimental controls in order to eliminate the influence of the third variable or using statistical methods to eliminate its influence. For example, one may wish to study the relation between memorization ability and ability to solve certain sorts of problems. Both of these skills may be related to intelligence; therefore in order to determine their direct relation to each other the influence of differences in intelligence must be controlled. To effect *experimental* control, we might choose subjects with equal intelligence. But if experimental controls are not feasible, then *statistical* controls can be applied. By the technique of partial correlation we could hold constant the effect of intelligence on the relation between memorization ability and ability to solve problems, and thereby determine the extent of the direct or uncontaminated relation between these two skills.

In this section we shall present a method of statistical control which may be used with the Kendall rank correlation τ. To use this nonparametric method of partial correlation, we must have data which are measured in at least an ordinal scale. No assumptions about the shape of the population of scores need be made.

Rationale

Suppose we obtain ranks of 4 subjects on 3 variables: X, Y, and Z. We wish to determine the correlation between X and Y when Z is partialled out (held constant). The ranks are

Subject	a	b	c	d
Rank on Z	1	2	3	4
Rank on X	3	1	2	4
Rank on Y	2	1	3	4

Now if we consider the possible pairs of ranks on any variable, we know that there are $\binom{4}{2}$ possible pairs—four things taken two at a time. Having arranged the ranks on Z in natural order, let us observe every possible pair in the X ranks, the Y ranks, and the Z ranks. We shall assign a $+$ to each of those pairs in which the lower rank precedes the higher, and a $-$ to each pair in which the higher rank precedes the lower:

Pair	(a,b)	(a,c)	(a,d)	(b,c)	(b,d)	(c,d)
Z	$+$	$+$	$+$	$+$	$+$	$+$
X	$-$	$-$	$+$	$+$	$+$	$+$
Y	$-$	$+$	$+$	$+$	$+$	$+$

That is, for variable X the score for the pair (a,b) is a $-$ because the ranks for a and b, 3 and 1, occur in the "wrong" order—the higher rank precedes the lower. For variable X, the score for the pair (a,c) is also a $-$ because the a rank, 3, is higher than the c rank, 2. For variable Y, the pair (a,c) receives a $+$ because the a rank, 2, is lower than the c rank, 3.

Now we may summarize the information we have obtained by casting it in a 2 × 2 table, Table 9.8.   Consider first the three signs under (a,b) above.   For that set of paired ranks, both X and Y are assigned a −, whereas Z is assigned a +.   Thus we say that both X and Y "disagree" with Z.   We summarize that information by casting pair (a,b) in cell D of Table 9.8.   Consider next the pair (a,c).   Here Y's sign agrees with

TABLE 9.8

	Y pairs whose sign agrees with Z's sign	Y pairs whose sign disagrees with Z's sign	Total
X pairs whose sign agrees with Z's sign	A 4	B 0	4
X pairs whose sign disagrees with Z's sign	C 1	D 1	2
Total	5	1	6

Z's sign, but X's sign disagrees with Z's sign.   Therefore pair (a,c) is assigned to cell C in Table 9.8.   In each case of the remaining pairs, both Y's sign and X's sign agree with Z's sign; thus these 4 pairs are cast in cell A of Table 9.8.

TABLE 9.9. FORM FOR CASTING DATA FOR COMPUTATION BY FORMULA (9.12)

	Y pairs whose sign agrees with Z's sign	Y pairs whose sign disagrees with Z's sign	Total
X pairs whose sign agrees with Z's sign	A	B	A + B
X pairs whose sign disagrees with Z's sign	C	D	C + D
Total	A + C	B + D	$\binom{N}{2}$

In general, for three sets of rankings of N objects, we can use the method illustrated above to derive the sort of table for which Table 9.9 is a model.   The Kendall partial rank correlation coefficient, $\tau_{xy.z}$ (read: the correlation between X and Y with Z held constant) is computed from such a table.   It is defined as

$$\tau_{xy.z} = \frac{AD - BC}{\sqrt{(A + B)(C + D)(A + C)(B + D)}} \qquad (9.12)$$

In the case of the 4 objects we have been considering, i.e., in the case of the data shown in Table 9.8,

$$\tau_{xy.z} = \frac{(4)(1) - (0)(1)}{\sqrt{(4)(2)(5)(1)}}$$

$$= .63$$

The correlation between X and Y with the effect of Z held constant is expressed by $\tau_{xy.z} = .63$. If we had computed the correlation between X and Y without considering the effect of Z, we would have found $\tau = .67$. This suggests that the relations between X and Z and between Y and Z are only slightly influencing the observed relation between X and Y. This kind of inference, however, must be made with reservations unless there are relevant prior grounds for expecting whatever effect is observed.

Formula (9.12) is sometimes called the "phi coefficient," and it can be shown that

$$\tau_{xy.z} = \sqrt{\frac{\chi^2}{N}}$$

The presence of χ^2 in the expression suggests that $\tau_{xy.z}$ measures the extent to which X and Y agree *independently* of their agreement with Z.

Method

Although the method which we have shown for computing $\tau_{xy.z}$ is useful in revealing the nature of the partial correlation, as N gets larger this method rapidly becomes more tedious because of the rapid increase of the value of $\binom{N}{2}$.

Kendall (1948a, p. 103) has shown that

$$\tau_{xy.z} = \frac{\tau_{xy} - \tau_{zy}\tau_{xz}}{\sqrt{(1 - \tau_{zy}^2)(1 - \tau_{zx}^2)}} \tag{9.13}*$$

Formula (9.13) is computationally easier than formula (9.12). To use it, one first must find the correlations (τ's) between X and Y, X and Z, and Y and Z. Having these values, one may use formula (9.13) to find $\tau_{xy.z}$.

For the X, Y, and Z ranks we have been considering, $\tau_{xy} = .67$, $\tau_{zy} = .67$, and $\tau_{xz} = .33$. Inserting these values in formula (9.13), we have

$$\tau_{xy.z} = \frac{.67 - (.67)(.33)}{\sqrt{[1 - (.67)^2][1 - (.33)^2]}}$$

$$= .63$$

---

* This formula is directly comparable to that used in finding the parametric partial product-moment correlation. Kendall (1948a, p. 103) states that the similarity seems to be merely coincidental.

Using formula (9.13), we arrive at the same value of $\tau_{xy.z}$ we have already arrived at by using formula (9.12).

Example

We have already seen that in the data collected by Siegel and Fagan, the correlation between scores on authoritarianism and scores on social status strivings is $\tau = .67$. However, we have also observed that there is a correlation between social status strivings and amount of conformity (yielding) to group pressures: $\tau = .39$. This may make us wonder whether the first-mentioned correlation

TABLE 9.10. RANKS ON AUTHORITARIANISM, SOCIAL STATUS STRIVINGS, AND CONFORMITY

Subject	Rank		
	Social status striving	Authoritarianism	Conformity (yielding)
	X	Y	Z
A	3	2	1.5
B	4	6	1.5
C	2	5	3.5
D	1	1	3.5
E	8	10	5.0
F	11	9	6.0
G	10	8	7.0
H	6	3	8.0
I	7	4	9.0
J	12	12	10.5
K	5	7	10.5
L	9	11	12.0

simply represents the operation of a third variable: conformity to group pressures. That is, it may be that the subjects' need to conform affects their responses to both the authoritarianism scale and the social status strivings scale, and thus the correlation between the scores on these two scales may be due to an association between each variable and need to conform. We may check whether this is true by computing a partial correlation between authoritarianism and social status strivings, partialling out the effect of need to conform, as indicated by amount of yielding in the Asch situation.

The scores for the 12 subjects on each of the three variables are shown in Tables 9.3 and 9.5. The three sets of ranks are shown in Table 9.10. Observe that the variable whose effect we wish to partial out, conformity, is the Z variable.

We have already determined that the correlation between social status strivings (the X variable) and authoritarianism (the Y variable) is $\tau_{xy} = .67$. We have also already determined that the correlation between social status strivings and conformity is $\tau_{xz} = .39$ (this value is corrected for ties). From the data presented in Table 9.10, we may readily determine, using formula (9.10), that the correlation between conformity and authoritarianism is $\tau_{zy} = .36$ (this value is corrected for ties). With that information, we may determine the value of $\tau_{xy.z}$ by using formula (9.13):

$$\tau_{xy.z} = \frac{\tau_{xy} - \tau_{zy}\tau_{xz}}{\sqrt{(1 - \tau_{zy}{}^2)(1 - \tau_{xz}{}^2)}} \tag{9.13}$$

$$= \frac{.67 - (.36)(.39)}{\sqrt{[1 - (.36)^2][1 - (.39)^2]}}$$

$$= .62$$

We have determined that when conformity is partialled out, the correlation between social status strivings and authoritarianism is $\tau_{xy.z} = .62$. Since this value is not much smaller than $\tau_{xy} = .67$, we might conclude that the relation between social status strivings and authoritarianism (as measured by these scales) is relatively independent of the influence of conformity (as measured in terms of amount of yielding to group pressures).

Summary of Procedure. These are the steps in the use of the Kendall partial rank correlation coefficient:

1. Let X and Y be the two variables whose relation is to be determined, and let Z be the variable whose effect on X and Y is to be partialled out or held constant.

2. Rank the observations on the X variable from 1 to N. Do the same for the observations on the Y and Z variables.

3. Using either formula (9.9) or formula (9.10) (the latter is to be used when ties have occurred in either of the variables being correlated), determine the observed values of τ_{xy}, τ_{zy}, and τ_{xz}.

4. With those values, compute the value of $\tau_{xy.z}$, using formula (9.13).

Test of Significance

Unfortunately, the sampling distribution of the Kendall partial rank correlation is not as yet known, and therefore no tests of the significance of an observed $\tau_{xy.z}$ are now possible. It might be thought that with

$$\tau_{xy.z} = \sqrt{\frac{\chi^2}{N}}$$

a χ^2 test could be used. This is not so because the entities in cells A,

B, C, and D of a table like Table 9.9 are not independent (their sum is $\binom{N}{2}$ rather than N) and a χ^2 test may properly and meaningfully be made only on independent observations.

References

The reader may find other discussions of this statistic in Kendall (1948a, chap. 8) and in Moran (1951).

THE KENDALL COEFFICIENT OF CONCORDANCE: W

Function

In the previous sections of this chapter, we have been concerned with measures of the correlation between *two* sets of rankings of N objects or individuals. Now we shall consider a measure of the relation among *several* rankings of N objects or individuals.

When we have k sets of rankings, we may determine the association among them by using the Kendall coefficient of concordance W. Whereas r_S and τ express the degree of association between two variables measured in, or transformed to, ranks, W expresses the degree of association among k such variables. Such a measure may be particularly useful in studies of interjudge or intertest reliability, and also has applications in studies of clusters of variables.

Rationale

As a solution to the problem of ascertaining the over-all agreement among k sets of rankings, it might seem reasonable to find the r_S's (or τ's) between all possible pairs of the rankings and then compute the average of these coefficients to determine the over-all association. In following such a procedure, we would need to compute $\binom{k}{2}$ rank correlation coefficients. Unless k were very small, such a procedure would be extremely tedious.

The computation of W is much simpler, and W bears a linear relation to the average r_S taken over all groups. If we denote the average value of the Spearman rank correlation coefficients between the $\binom{k}{2}$ possible pairs of rankings as $r_{S_{av}}$, then it has been shown (Kendall, 1948a, p. 81) that

$$r_{S_{av}} = \frac{kW - 1}{k - 1} \tag{9.14}$$

Another approach would be to imagine how our data would look if there were no agreement among the several sets of rankings, and then to imagine how it would look if there were perfect agreement among the several sets. The coefficient of concordance would then be an index of the divergence of the actual agreement shown in the data from the maximum possible (perfect) agreement. Very roughly speaking, W is just such a coefficient.

Suppose three company executives are asked to interview six job applicants and to rank them separately in their order of suitability for a job opening. The three independent sets of ranks given by executives X, Y, and Z to applicants a through f might be those shown in Table 9.11.

TABLE 9.11. RANKS ASSIGNED TO SIX JOB APPLICANTS BY THREE
COMPANY EXECUTIVES
(Artificial data)

	Applicant					
	a	b	c	d	e	f
Executive X	1	6	3	2	5	4
Executive Y	1	5	6	4	2	3
Executive Z	6	3	2	5	4	1
R_j	8	14	11	11	11	8

The bottom row of Table 9.11, labeled R_j, gives the sums of the ranks assigned to each applicant.

Now if the three executives had been in *perfect* agreement about the applicants, i.e., if they had each ranked the six applicants in the same order, then one applicant would have received three ranks of 1 and thus his sum of ranks, R_j, would be $1 + 1 + 1 = 3 = k$. The applicant whom all executives designated as the runner-up would have

$$R_j = 2 + 2 + 2 = 6 = 2k$$

The least promising applicant would have

$$R_j = 6 + 6 + 6 = 18 = Nk$$

In fact, with perfect agreement among the executives, the various sums of ranks, R_j, would be these: 3, 6, 9, 12, 15, 18, though not necessarily in that order. In general, when there is perfect agreement among k sets of rankings, we get, for the R_j, the series: k, $2k$, $3k$, . . . , Nk.

On the other hand, if there had been no agreement among the three executives, then the various R_j's would be approximately equal.

From this example, if should be clear that the degree of agreement among the k judges is reflected by the degree of variance among the N sums of ranks. W, the coefficient of concordance, is a function of that degree of variance.

Method

To compute W, we first find the sum of ranks, R_j, in each column of a $k \times N$ table. Then we sum the R_j and divide that sum by N to obtain the mean value of the R_j. Each of the R_j may then be expressed as a deviation from the mean value. (We have shown above that the larger are these deviations, the greater is the degree of association among the k sets of ranks.) Finally, s, the sum of squares of these deviations, is found. Knowing these values, we may compute the value of W:

$$W = \frac{s}{\frac{1}{12}k^2(N^3 - N)} \tag{9.15}$$

where s = sum of squares of the observed deviations from the mean of R_j, that is, $s = \sum \left(R_j - \frac{\Sigma R_j}{N} \right)^2$

k = number of sets of rankings, e.g., the number of judges
N = number of entities (objects or individuals) ranked
$\frac{1}{12}k^2(N^3 - N)$ = maximum possible sum of the squared deviations, i.e., the sum s which would occur with perfect agreement among k rankings

For the data shown in Table 9.11, the rank totals were 8, 14, 11, 11, 11, and 8. The mean of these values is 10.5. To obtain s, we square the deviation of each rank total from that mean value, and then sum those squares:

$$s = (8 - 10.5)^2 + (14 - 10.5)^2 + (11 - 10.5)^2 + (11 - 10.5)^2$$
$$+ (11 - 10.5)^2 + (8 - 10.5)^2$$
$$= 25.5$$

Knowing the observed value of s, we may find the value of W for the data in Table 9.11 by using formula (9.15):

$$W = \frac{25.5}{\frac{1}{12}(3)^2(6^3 - 6)}$$
$$= .16$$

$W = .16$ expresses the degree of agreement among the three fictitious executives in ranking the six job applicants.

With the same data, we might have found $r_{S_{av}}$ by either of two methods. One way would be first to find the values of $r_{S_{xy}}$, $r_{S_{xz}}$, and $r_{S_{yz}}$. Then these three values could be averaged. For the data in Table 9.11, $r_{s_{xy}} = .31$, $r_{s_{yz}} = -.54$, and $r_{s_{xz}} = -.54$. The average of these values is

$$r_{S_{av}} = \frac{.31 + (-.54) + (-.54)}{3}$$

$$= -.26$$

Another way to find $r_{S_{av}}$ would be to use formula (9.14):

$$r_{S_{av}} = \frac{kW - 1}{k - 1} \tag{9.14}$$

$$= \frac{3(.16) - 1}{3 - 1}$$

$$= -.26$$

Both methods yield the same value: $r_{S_{av}} = -.26$. As is shown above, this value bears a linear relation to the value of W.

One difference between the W and the $r_{S_{av}}$ methods of expressing agreement among k rankings is that $r_{S_{av}}$ may take values between -1 and $+1$, whereas W may take values only between 0 and $+1$. The reason that W cannot be negative is that when more than two sets of ranks are involved, the rankings cannot all disagree completely. For example, if judge X and judge Y are in disagreement, and judge X is also in disagreement with judge Z, then judges Y and Z must agree. That is, when more than two judges are involved, agreement and disagreement are not symmetrical opposites. k judges may all agree, but they cannot all disagree completely. Therefore W must be zero or positive.

The reader should notice that W bears a linear relation to r_S but seems to bear no orderly relation to τ. This reveals one of the advantages which r_S has over τ.

Example

Twenty mothers and their deaf preschool children attended a summer camp designed to give introductory training in the treatment and handling of deaf children. A staff of 13 psychologists and speech correctionists worked with the mothers and children during the 2-week camp session. At the end of that period, the 13 staff members were asked to rank the 20 mothers on how likely it was that each mother would rear her child in such a way that the child would suffer personal maladjustment.[1] These rankings are shown in Table 9.12.

---

[1] This example cites unpublished data from research conducted at the 1955 Camp Easter Seal Speech and Hearing Program, Laurel Hill State Park, Pa. The data were made available to the author through the courtesy of the researcher, Dr. J. E. Gordon.

A coefficient of concordance was computed to determine the agreement among the staff members. The mean of the various R_j is 135.5. The deviation of every R_j from that mean, and the square of that deviation, are shown in Table 9.12. The sum of these squares $= 64,899 = s$. $k = 13 =$ the number of judges. $N = 20 =$ the

TABLE 9.12. RANKS ASSIGNED TO 20 MOTHERS BY 13 STAFF MEMBERS

Judge	Mother 1	2	3	4	5	6	7	8	9	10	11	12	13	14	15	16	17	18	19	20
A	1	2	3	4	5	6	7	8	9	10	11	12	13	14	15	16	17	18	19	20
B	5	1	16	8	9	2	6	10	4	3	11	13	7	12	17	18	19	15	14	20
C	3	2	7	5	14	9	15	16	6	11	8	10	1	4	19	12	20	13	17	18
D	8	3	10	11	4	2	5	13	9	1	14	7	6	15	16	12	19	17	18	20
E	2	1	15	8	14	4	6	9	7	10	11	5	3	16	11	13	18	17	12	19
F	16	17	5	13	15	11	7	4	9	2	18	3	6	1	19	12	10	8	14	20
G	12	9	14	6	7	2	3	10	5	4	17	8	1	15	13	16	18	11	20	19
H	11	2	13	10	7	3	4	14	6	5	17	9	1	12	8	16	20	15	18	19
I	9	2	15	6	5	7	8	10	9	3	12	4	1	13	11	14	19	18	16	17
J	2	4	16	3	10	6	14	17	15	7	19	9	1	8	5	13	11	18	12	20
K	11	14	12	8	7	2	5	10	3	4	13	9	1	18	6	15	19	16	17	20
L	8	1	13	3	5	2	14	9	6	10	15	11	19	4	7	12	18	17	16	20
M	5	3	13	2	8	1	9	12	4	6	14	10	11	7	15	18	16	17	19	20
R_j	93	61	152	87	110	57	103	142	92	76	180	110	71	139	162	187	224	200	212	252
$R_j - \dfrac{\Sigma R_j}{N}$	−42.5	−74.5	16.5	−48.5	−25.5	−78.5	−32.5	6.5	−43.5	−59.5	44.5	−25.5	−64.5	3.5	26.5	51.5	88.5	64.5	76.5	116.5
$\left(R_j - \dfrac{\Sigma R_j}{N}\right)^2$	1,806.25	5,550.25	272.25	2,352.25	650.25	6,162.25	1,056.25	42.25	1,892.25	3,540.25	1,980.25	650.25	4,160.25	12.25	702.25	2,652.25	7,832.25	4,160.25	5,852.25	13,572.25

number of mothers who were ranked. With this information, we may compute W:

$$W = \frac{s}{\frac{1}{12}k^2(N^3 - N)} \tag{9.15}$$

$$= \frac{64,899}{\frac{1}{12}(13)^2[(20)^3 - 20]}$$

$$= .577$$

The agreement among the 13 staff members is expressed by

$$W = .577$$

Tied observations. When tied observations occur, the observations are each assigned the average of the ranks they would have been assigned had no ties occurred, our usual procedure in ranking tied scores.

The effect of tied ranks is to depress the value of W as found by formula (9.15). If the proportion of ties is small, that effect is negligible, and thus formula (9.15) may still be used. If the proportion of ties is large, a correction may be introduced which will increase slightly the value of W over what it would have been if uncorrected. That correction factor is the same one used with the Spearman r_S:

$$T = \frac{\Sigma(t^3 - t)}{12}$$

where t = number of observations in a group tied for a given rank

Σ directs one to sum over all groups of ties within any one of the k rankings

With the correction of ties incorporated, the Kendall coefficient of concordance is

$$W = \frac{s}{\frac{1}{12}k^2(N^3 - N) - k \sum_{T} T} \qquad (9.16)$$

where $\sum_{T} T$ directs one to sum the values of T for all the k rankings.

Example with Ties

Kendall (1948a, p. 83) has given an example in which 10 objects are each ranked on 3 different variables: X, Y, and Z. The ranks are shown in Table 9.13, which also shows the values of R_j.

TABLE 9.13. RANKS RECEIVED BY TEN ENTITIES ON THREE VARIABLES

Variable	Entity									
	a	b	c	d	e	f	g	h	i	j
X	1	4.5	2	4.5	3	7.5	6	9	7.5	10
Y	2.5	1	2.5	4.5	4.5	8	9	6.5	10	6.5
Z	2	1	4.5	4.5	4.5	4.5	8	8	8	10
R_j	5.5	6.5	9	13.5	12	20	23	23.5	25.5	26.5

The mean of the R_j is 16.5. To obtain s, we sum the squared deviations of each R_j from this mean:

$$s = (5.5 - 16.5)^2 + (6.5 - 16.5)^2 + (9 - 16.5)^2 + (13.5 - 16.5)^2$$
$$+ (12 - 16.5)^2 + (20 - 16.5)^2 + (23 - 16.5)^2$$
$$+ (23.5 - 16.5)^2 + (25.5 - 16.5)^2 + (26.5 - 16.5)^2$$
$$= 591$$

Since the proportion of ties in the ranks is large, we should correct for ties in computing the value of W.

In the X rankings, there are two sets of ties: 2 objects are tied at 4.5 and 2 are tied at 7.5. For both groups, t = the number of observations tied for a given rank = 2. Thus

$$T_X = \frac{\Sigma(t^3 - t)}{12} = \frac{(2^3 - 2) + (2^3 - 2)}{12} = 1$$

In the Y rankings, there are three sets of ties, and each set contains two observations. Here $t = 2$ in each case, and

$$T_Y = \frac{\Sigma(t^3 - t)}{12} = \frac{(2^3 - 2) + (2^3 - 2) + (2^3 - 2)}{12} = 1.5$$

In the Z rankings, there are two sets of ties. One set, tied at 4.5, consists of 4 observations: here $t = 4$. The other set, tied at rank 8, consists of 3 observations: $t = 3$. Thus

$$T_Z = \frac{\Sigma(t^3 - t)}{12} = \frac{(4^3 - 4) + (3^3 - 3)}{12} = 7$$

Knowing the values of T for the X, Y, and Z rankings, we may find their sum: $\sum_T T = 1 + 1.5 + 7 = 9.5$.

With the above information, we may compute W corrected for ties:

$$W = \frac{s}{\frac{1}{12}k^2(N^3 - N) - k\sum_T T} \tag{9.16}$$

$$= \frac{591}{\frac{1}{12}(3)^2[(10)^3 - 10] - 3(9.5)}$$

$$= .828$$

If we had disregarded the ties, i.e., if we had used formula (9.15) in computing W, we would have found $W = .796$ rather than $W = .828$. This difference illustrates the slightly depressing effect which ties, when uncorrected, exert on the value of W.

Testing the Significance of W

Small samples. We may test the significance of any observed value of W by determining the probability associated with the occurrence under H_0 of a value as large as the s with which it is associated. If we obtain the sampling distribution of s for all permutations in the N ranks in all possible ways in the k rankings, we will have $(N!)^k$ sets of possible ranks.

Using these, we may test the null hypothesis that the k sets of rankings are independent by taking from this distribution the probability associated with the occurrence under H_0 of a value as large as an observed s.

By this method, the distribution of s under H_0 has been worked out and certain critical values have been tabled. Table R of the Appendix gives values of s for W's significant at the .05 and .01 levels. This table is applicable for k from 3 to 20, and for N from 3 to 7. If an observed s is equal to or greater than that shown in Table R for a particular level of significance, then H_0 may be rejected at that level of significance.

For example, we saw that when $k = 3$ fictitious executives ranked $N = 6$ job applicants, their agreement was $W = .16$. Reference to Table R reveals that the s associated with that value of W ($s = 25.5$) is not significant. For the association to have been significant at the .05 level, s would have had to be 103.9 or larger.

Large samples. When N is larger than 7, the expression given in formula (9.17) is approximately distributed as chi square with

$$df = N - 1$$

$$\chi^2 = \frac{s}{\frac{1}{12}kN(N + 1)} \tag{9.17}$$

That is, the probability associated with the occurrence under H_0 of any value as large as an observed W may be determined by finding χ^2 by formula (9.17) and then determining the probability associated with so large a value of χ^2 by referring to table C of the Appendix.

Observe that

$$\frac{s}{\frac{1}{12}kN(N + 1)} = k(N - 1)W$$

and therefore $$\chi^2 = k(N - 1)W \tag{9.18}$$

Thus one may use formula (9.18), which is computationally simpler than formula (9.17), with $df = N - 1$, to determine the probability associated with the occurrence under H_0 of any value as large as an observed W.

If the value of χ^2 as computed from formula (9.18) [or, equivalently, from formula (9.17)] equals or exceeds that shown in Table C for a particular level of significance and a particular value of $df = N - 1$, then the null hypothesis that the k rankings are unrelated may be rejected at that level of significance.

Example[1]

In the study of ratings by staff persons of the mother-child relations of 20 mothers with their deaf young children, $k = 13$, $N = 20$,

[1] See footnote, page 211.

and we found that $W = .577$. We may determine the significance of this relation by applying formula (9.18):

$$\chi^2 = k(N - 1)W \qquad (9.18)$$
$$= 13(20 - 1)(.577)$$
$$= 142.5$$

Referring to Table C, we find that $\chi^2 \geq 142.5$ with

$$df = N - 1 = 20 - 1 = 19$$

has probability of occurrence under H_0 of $p < .001$. We can conclude with considerable assurance that the agreement among the 13 judges is higher than it would be by chance. The very low probability under H_0 associated with the observed value of W enables us to reject the null hypothesis that the judges' ratings are unrelated to each other.

Summary of Procedure

These are the steps in the use of W, the Kendall coefficient of concordance:

1. Let $N =$ the number of entities to be ranked, and let $k =$ the number of judges assigning ranks. Cast the observed ranks in a $k \times N$ table.

2. For each entity, determine R_j, the sum of the ranks assigned to that entity by the k judges.

3. Determine the mean of the R_j. Express each R_j as a deviation from that mean. Square these deviations, and sum the squares to obtain s.

·4. If the proportion of ties in the k sets of ranks is large, use formula (9.16) in computing the value of W. Otherwise use formula (9.15).

5. The method for determining whether the observed value of W is significantly different from zero depends on the size of N:

    a. If N is 7 or smaller, Table R gives critical values of s associated with W's significant at the .05 and .01 levels.

    b. If N is larger than 7, either formula (9.17) or formula (9.18) (the latter is easier) may be used to compute a value of χ^2 whose significance, for $df = N - 1$, may be tested by reference to Table C.

Interpretation of W

A high or significant value of W may be interpreted as meaning that the observers or judges are applying essentially the same standard in ranking the N objects under study. Often their pooled ordering may serve as a "standard," especially when there is no relevant external criterion for ordering the objects.

It should be emphasized that a high or significant value of W does *not* mean that the orderings observed are *correct*. In fact, they may all be incorrect with respect to some external criterion. For example, the 13 staff members of the camp agreed well in judging which mothers and their children were headed for difficulty, but only time can show whether their judgments were sound. It is possible that a variety of judges can agree in ordering objects because all employ the "wrong" criterion. In this case, a high or significant W would simply show that all more or less agree in their use of a "wrong" criterion. To state the point another way, a high degree of agreement about an order does not necessarily mean that the order which was agreed upon is the "objective" one. In the behavioral sciences, especially in psychology, "objective" orderings and "consensual" orderings are often incorrectly thought to be synonymous.

Kendall (1948a, p. 87) suggests that the best estimate of the "true" ranking of the N objects is provided, when W is significant, by the order of the various sums of ranks, R_j. If one accepts the criterion which the various judges have agreed upon (as evidenced by the magnitude and significance of W) in ranking the N entities, then the best estimate of the "true" ranking of those entities according to that criterion is provided by the order of the sums of ranks. This "best estimate" is associated, in a certain sense, with least squares. Thus our best estimate would be that either applicant a or f (see Table 9.11) should be hired for the job opening, for in both of their cases $R_j = 8$, the lowest value observed. And our best estimate would be that, of the 20 mothers of the deaf children, mother 6 (see Table 9.12), whose $R = 57$ is the smallest of the R_j, is the mother who is most likely to rear a well-adjusted child. Mother 2 is the next most likely, and mother 20 is the mother who, by consensus, is the one most likely to rear a maladjusted child.

References

Discussions of the Kendall coefficient of concordance are contained in Friedman (1940), Kendall (1948a, chap. 6), and Willerman (1955).

DISCUSSION

In this chapter we have presented five nonparametric techniques for measuring the degree of correlation between variables in a sample. For each of these, except the Kendall partial correlation coefficient, tests of the significance of the observed association were presented.

One of these techniques, the coefficient of contingency, is uniquely applicable when the data are in a nominal scale. That is, if the measurement is so crude that the classifications involved are unrelated within any set and thus cannot be meaningfully ordered, then the contingency coefficient is a meaningful measure of the degree of association in the data. For other suitable measures, see Kruskal and Goodman (1954). If the variables under study have been measured in at least an ordinal scale, the contingency coefficient may still be used, but an appropriate method of *rank* correlation will utilize more of the information in the data and therefore is preferable.

For the bivariate case two rank correlation coefficients, the Spearman r_S and the Kendall τ, were presented. The Spearman r_S is somewhat easier to compute, and has the further advantage of being linearly related to the coefficient of concordance W. However, the Kendall τ has the advantages of being generalizable to a partial correlation coefficient and of having a sampling distribution which is practically indistinguishable from a normal distribution for sample sizes as small as 9.

Both r_S and τ have the same power-efficiency (91 per cent) in testing for the existence of a relation in the population. That is, with data which meet the assumptions of the Pearson r, both r_S and τ are as powerful as r for rejecting the null hypothesis when r_S and τ are based on 10 observations for every 9 observations used in computing r.

The Kendall partial rank correlation coefficient measures the degree of relation between two variables, X and Y, when a third variable, Z (on which the association between X and Y might logically depend), is held constant. $\tau_{xy.z}$ is the nonparametric equivalent of the partial product moment r. However, no test of the significance of partial τ is as yet available.

The Kendall coefficient of concordance W measures the extent of association among several (k) sets of rankings of N entities. It is useful in determining the agreement among several judges or the association among three or more variables. It has special applications in providing a standard method of ordering entities according to consensus when there is available no objective order of the entities.

REFERENCES

Anderson, R. L., and Bancroft, T. A. 1952. *Statistical theory in research.* New York: McGraw-Hill.

Andrews, F. C. 1954. Asymptotic behavior of some rank tests for analysis of variance. *Ann. Math. Statist.,* **25,** 724–736.

Auble, D. 1953. Extended tables for the Mann-Whitney statistic. *Bull. Inst. Educ. Res. Indiana Univer.,* **1,** No. 2.

Barnard, G. A. 1947. Significance tests for 2 × 2 tables. *Biometrika,* **34,** 123–138.

Bergman, G., and Spence, K. W. 1944. The logic of psychological measurement. *Psychol. Rev.,* **51,** 1–24.

Birnbaum, Z. W. 1952. Numerical tabulation of the distribution of Kolmogorov's statistic for finite sample values. *J. Amer. Statist. Ass.,* **47,** 425–441.

Birnbaum, Z. W. 1953. Distribution-free tests of fit for continuous distribution functions. *Ann. Math. Statist.,* **24,** 1–8.

Birnbaum, Z. W., and Tingey, F. H. 1951. One-sided confidence contours for probability distribution functions. *Ann. Math. Statist.,* **22,** 592–596.

Blackwell, D., and Girshick, M. A. 1954. *Theory of games and statistical decisions.* New York: Wiley.

Blum, J. R., and Fattu, N. A. 1954. Nonparametric methods. *Rev. Educ. Res.,* **24,** 467–487.

Bowker, A. H. 1948. A test for symmetry in contingency tables. *J. Amer. Statist. Ass.,* **43,** 572–574.

Brown, G. W., and Mood, A. M. 1951. On median tests for linear hypotheses. *Proceedings of the second Berkeley symposium on mathematical statistics and probability.* Berkeley, Calif.: Univer. of Calif. Press. Pp. 159–166.

Clopper, C. J., and Pearson, E. S. 1934. The use of confidence or fiducial limits illustrated in the case of the binomial. *Biometrika,* **26,** 404–413.

Cochran, W. G. 1950. The comparison of percentages in matched samples. *Biometrika,* **37,** 256–266.

Cochran, W. G. 1952. The χ^2 test of goodness of fit. *Ann. Math. Statist.,* **23,** 315–345.

Cochran, W. G. 1954. Some methods for strengthening the common χ^2 tests. *Biometrics,* **10,** 417–451.

Coombs, C. H. 1950. Psychological scaling without a unit of measurement. *Psychol. Rev.,* **57,** 145–158.

Coombs, C. H. 1952. A theory of psychological scaling. *Bull. Univer. Michigan Engng Res. Inst.,* **34.**

David, F. N. 1949. *Probability theory for statistical methods.* New York: Cambridge Univer. Press.

Davidson, D., Siegel, S., and Suppes, P. 1955. *Some experiments and related theory*
241

on the measurement of utility and subjective probability. Rep. 4, Stanford Value Theory Project.

Dixon, W. J. 1954. Power under normality of several non-parametric tests. *Ann. Math. Statist.*, **25**, 610–614.

Dixon, W. J., and Massey, F. J. 1951. *Introduction to statistical analysis.* New York: McGraw-Hill.

Dixon, W. J., and Mood, A. M. 1946. The statistical sign test. *J. Amer. Statist. Ass.*, **41**, 557–566.

Edwards, A. L. 1954. *Statistical methods for the behavioral sciences.* New York: Rinehart.

Festinger, L. 1946. The significance of differences between means without reference to the frequency distribution function. *Psychometrika*, **11**, 97–105.

Finney, D. J. 1948. The Fisher-Yates test of significance in 2 × 2 contingency tables. *Biometrika*, **35**, 145–156.

Fisher, R. A. 1934. *Statistical methods for research workers.* (5th Ed.) Edinburgh: Oliver & Boyd.

Fisher, R. A. 1935. *The design of experiments.* Edinburgh: Oliver & Boyd.

Freund, J. E. 1952. *Modern elementary statistics.* New York: Prentice-Hall.

Friedman, M. 1937. The use of ranks to avoid the assumption of normality implicit in the analysis of variance. *J. Amer. Statist. Ass.*, **32**, 675–701.

Friedman, M. 1940. A comparison of alternative tests of significance for the problem of *m* rankings. *Ann. Math. Statist.*, **11**, 86–92.

Goodman, L. A. 1954. Kolmogorov-Smirnov tests for psychological research. *Psychol. Bull.*, **51**, 160–168.

Goodman, L. A., and Kruskal, W. H. 1954. Measures of association for cross classifications. *J. Amer. Statist. Ass.*, **49**, 732–764.

Hempel, C. G. 1952. Fundamentals of concept formation in empirical science. *Int. Encycl. Unif. Sci.*, **2**, No. 7. (Univer. of Chicago Press.)

Hotelling, H., and Pabst, Margaret R. 1936. Rank correlation and tests of significance involving no assumption of normality. *Ann. Math. Statist.*, **7**, 29–43.

Jonckheere, A. R. 1954. A distribution-free k-sample test against ordered alternatives. *Biometrika*, **41**, 133–145.

Kendall, M. G. 1938. A new measure of rank correlation. *Biometrika*, **30**, 81–93.

Kendall, M. G. 1945. The treatment of ties in ranking problems. *Biometrika*, **33**, 239–251.

Kendall, M. G. 1947. The variance of τ when both rankings contain ties. *Biometrika*, **34**, 297–298.

Kendall, M. G. 1948a. *Rank correlation methods.* London: Griffin.

Kendall, M. G. 1948b. *The advanced theory of statistics.* Vol. 1. (4th Ed.) London: Griffin.

Kendall, M. G. 1949. Rank and product-moment correlation. *Biometrika*, **36**, 177–193.

Kendall, M. G., and Smith, B. B. 1939. The problem of *m* rankings. *Ann. Math. Statist.*, **10**, 275–287.

Kolmogorov, A. 1941. Confidence limits for an unknown distribution functon. *Ann. Math. Statist.*, **12**, 461–463.

Kruskal, W. H. 1952. A nonparametric test for the several sample problem. *Ann. Math. Statist.*, **23**, 525–540.

Kruskal, W. H., and Wallis, W. A. 1952. Use of ranks in one-criterion variance analysis. *J. Amer. Statist. Ass.*, **47**, 583–621.

Latscha, R. 1953. Tests of significance in a 2 × 2 contingency table: Extension of Finney's table. *Biometrika*, **40**, 74–86.

Lehmann, E. L. 1953. The power of rank tests. *Ann. Math. Statist.*, **24**, 23–43.

Lewis, D., and Burke, C. J. 1949. The use and misuse of the chi-square test. *Psychol. Bull.*, **46**, 433–489.

McNemar, Q. 1946. Opinion-attitude methodology. *Psychol. Bull.*, **43**, 289–374.

McNemar, Q. 1947. Note on the sampling error of the difference between correlated proportions or percentages. *Psychometrika*, **12**, 153–157.

McNemar, Q. 1955. *Psychological statistics.* (2nd Ed.) New York: Wiley.

Mann, H. B., and Whitney, D. R. 1947. On a test of whether one of two random variables is stochastically larger than the other. *Ann. Math. Statist.*, **18**, 50–60.

Massey, F. J., Jr. 1951a. The Kolmogorov-Smirnov test for goodness of fit. *J. Amer. Statist. Ass.*, **46**, 68–78.

Massey, F. J., Jr. 1951b. The distribution of the maximum deviation between two sample cumulative step functions. *Ann. Math. Statist.*, **22**, 125–128.

Mood, A. M. 1940. The distribution theory of runs. *Ann. Math. Statist.*, **11**, 367–392.

Mood, A. M. 1950. *Introduction to the theory of statistics.* New York: McGraw-Hill.

Mood, A. M. 1954. On the asymptotic efficiency of certain non-parametric two-sample tests. *Ann. Math. Statist.*, **25**, 514–522.

Moore, G. H., and Wallis, W. A. 1943. Time series significance tests based on signs of differences. *J. Amer. Statist. Ass.*, **38**, 153–164.

Moran, P. A. P. 1951. Partial and multiple rank correlation. *Biometrika*, **38**, 26–32.

Moses, L. E. 1952a. Non-parametric statistics for psychological research. *Psychol. Bull.*, **49**, 122–143.

Moses, L. E. 1952b. A two-sample test. *Psychometrika*, **17**, 239–247.

Mosteller, F. 1948. A k-sample slippage test for an extreme population. *Ann. Math. Statist.*, **19**, 58–65.

Mosteller, F., and Bush, R. R. 1954. Selected quantitative techniques. In G. Lindzey (Ed.), *Handbook of social psychology.* Vol. 1. *Theory and method.* Cambridge, Mass.: Addison-Wesley. Pp. 289–334.

Mosteller, F., and Tukey, J. W. 1950. Significance levels for a k-sample slippage test. *Ann. Math. Statist.*, **21**, 120–123.

Olds, E. G. 1949. The 5% significance levels for sums of squares of rank differences and a correction. *Ann. Math. Statist.*, **20**, 117–118.

Pitman, E. J. G. 1937a. Significance tests which may be applied to samples from any populations. Supplement to *J. R. Statist. Soc.*, **4**, 119–130.

Pitman, E. J. G. 1937b. Significance tests which may be applied to samples from any populations. II. The correlation coefficient test. Supplement to *J. R. Statist. Soc.*, **4**, 225–232.

Pitman, E. J. G. 1937c. Significance tests which may be applied to samples from any populations. III. The analysis of variance test. *Biometrika*, **29**, 322–335.

Savage, I. R. 1953. Bibliography of nonparametric statistics and related topics. *J. Amer. Statist. Ass.*, **48**, 844–906.

Savage, L. J. 1954. *The foundations of statistics.* New York: Wiley.

Scheffé, H. 1943. Statistical inference in the non-parametric case. *Ann. Math. Statist.*, **14**, 305–332.

Siegel, S. 1956. A method for obtaining an ordered metric scale. *Psychometrika*, **21**, 207–216.

Smirnov, N. V. 1948. Table for estimating the goodness of fit of empirical distributions. *Ann. Math. Statist.*, **19**, 279–281.

Smith, K. 1953. Distribution-free statistical methods and the concept of power efficiency. In L. Festinger and D. Katz (Eds.), *Research methods in the behavioral sciences*. New York: Dryden. Pp. 536–577.

Snedecor, G. W. 1946. *Statistical methods*. (4th Ed.) Ames, Iowa: Iowa State College Press.

Stevens, S. S. 1946. On the theory of scales of measurement. *Science*, 103, 677–680.

Stevens, S. S. 1951. Mathematics, measurement, and psychophysics. In S. S. Stevens (Ed.), *Handbook of experimental psychology*. New York: Wiley. Pp. 1–49.

Stevens, W. L. 1939. Distribution of groups in a sequence of alternatives. *Ann. Eugenics*, 9, 10–17.

Swed, Frieda S., and Eisenhart, C. 1943. Tables for testing randomness of grouping in a sequence of alternatives. *Ann. Math. Statist.*, 14, 66–87.

Tocher, K. D. 1950. Extension of the Neyman-Pearson theory of tests to discontinuous variates. *Biometrika*, 37, 130–144.

Tukey, J. W. 1949. Comparing individual means in the analysis of variance. *Biometrics*, 5, 99–114.

Wald, A. 1950. *Statistical decision functions*. New York: Wiley.

Walker, Helen M., and Lev, J. 1953. *Statistical inference*. New York: Holt.

Walsh, J. E. 1946. On the power function of the sign test for slippage of means. *Ann. Math. Statist.*, 17, 358–362.

Walsh, J. E. 1949a. Some significance tests for the median which are valid under very general conditions. *Ann. Math. Statist.*, 20, 64–81.

Walsh, J. E. 1949b. Applications of some significance tests for the median which are valid under very general conditions. *J. Amer. Statist. Ass.*, 44, 342–355.

Welch, B. L. 1937. On the z-test in randomized blocks and Latin squares. *Biometrika*, 29, 21–52.

White, C. 1952. The use of ranks in a test of significance for comparing two treatments. *Biometrics*, 8, 33–41.

Whitney, D. R. 1948. A comparison of the power of non-parametric tests and tests based on the normal distribution under non-normal alternatives. Unpublished doctor's dissertation, Ohio State Univer.

Whitney, D. R. 1951. A bivariate extension of the U statistic. *Ann. Math. Statist.*, 22, 274–282.

Wilcoxon, F. 1945. Individual comparisons by ranking methods. *Biometrics Bull.*, 1, 80–83.

Wilcoxon, F. 1947. Probability tables for individual comparisons by ranking methods. *Biometrics*, 3, 119–122.

Wilcoxon, F. 1949. *Some rapid approximate statistical procedures*. Stamford, Conn.: American Cyanamid Co.

Wilks, S. S. 1948. Order statistics. *Bull. Amer. Math. Soc.*, 54, 6–50

Willerman, B. 1955. The adaptation and use of Kendall's coefficient of concordance (W) to sociometric-type rankings. *Psychol. Bull.*, 52, 132-133.

Yates, F. 1934. Contingency tables involving small numbers and the χ^2 test. Supplement to *J. R. Statist. Soc.*, 1, 217-235.

APPENDIX

LIST OF TABLES

TABLE A. TABLE OF PROBABILITIES ASSOCIATED WITH VALUES AS EXTREME AS
OBSERVED VALUES OF z IN THE NORMAL DISTRIBUTION

The body of the table gives one-tailed probabilities under H_0 of z. The left-hand
marginal column gives various values of z to one decimal place. The top row gives
various values to the second decimal place. Thus, for example, the one-tailed p of
$z \geq .11$ or $z \leq -.11$ is $p = .4562$.

z	.00	.01	.02	.03	.04	.05	.06	.07	.08	.09
.0	.5000	.4960	.4920	.4880	.4840	.4801	.4761	.4721	.4681	.4641
.1	.4602	.4562	.4522	.4483	.4443	.4404	.4364	.4325	.4286	.4247
.2	.4207	.4168	.4129	.4090	.4052	.4013	.3974	.3936	.3897	.3859
.3	.3821	.3783	.3745	.3707	.3669	.3632	.3594	.3557	.3520	.3483
.4	.3446	.3409	.3372	.3336	.3300	.3264	.3228	.3192	.3156	.3121
.5	.3085	.3050	.3015	.2981	.2946	.2912	.2877	.2843	.2810	.2776
.6	.2743	.2709	.2676	.2643	.2611	.2578	.2546	.2514	.2483	.2451
.7	.2420	.2389	.2358	.2327	.2296	.2266	.2236	.2206	.2177	.2148
.8	.2119	.2090	.2061	.2033	.2005	.1977	.1949	.1922	.1894	.1867
.9	.1841	.1814	.1788	.1762	.1736	.1711	.1685	.1660	.1635	.1611
1.0	.1587	.1562	.1539	.1515	.1492	.1469	.1446	.1423	.1401	.1379
1.1	.1357	.1335	.1314	.1292	.1271	.1251	.1230	.1210	.1190	.1170
1.2	.1151	.1131	.1112	.1093	.1075	.1056	.1038	.1020	.1003	.0985
1.3	.0968	.0951	.0934	.0918	.0901	.0885	.0869	.0853	.0838	.0823
1.4	.0808	.0793	.0778	.0764	.0749	.0735	.0721	.0708	.0694	.0681
1.5	.0668	.0655	.0643	.0630	.0618	.0606	.0594	.0582	.0571	.0559
1.6	.0548	.0537	.0526	.0516	.0505	.0495	.0485	.0475	.0465	.0455
1.7	.0446	.0436	.0427	.0418	.0409	.0401	.0392	.0384	.0375	.0367
1.8	.0359	.0351	.0344	.0336	.0329	.0322	.0314	.0307	.0301	.0294
1.9	.0287	.0281	.0274	.0268	.0262	.0256	.0250	.0244	.0239	.0233
2.0	.0228	.0222	.0217	.0212	.0207	.0202	.0197	.0192	.0188	.0183
2.1	.0179	.0174	.0170	.0166	.0162	.0158	.0154	.0150	.0146	.0143
2.2	.0139	.0136	.0132	.0129	.0125	.0122	.0119	.0116	.0113	.0110
2.3	.0107	.0104	.0102	.0099	.0096	.0094	.0091	.0089	.0087	.0084
2.4	.0082	.0080	.0078	.0075	.0073	.0071	.0069	.0068	.0066	.0064
2.5	.0062	.0060	.0059	.0057	.0055	.0054	.0052	.0051	.0049	.0048
2.6	.0047	.0045	.0044	.0043	.0041	.0040	.0039	.0038	.0037	.0036
2.7	.0035	.0034	.0033	.0032	.0031	.0030	.0029	.0028	.0027	.0026
2.8	.0026	.0025	.0024	.0023	.0023	.0022	.0021	.0021	.0020	.0019
2.9	.0019	.0018	.0018	.0017	.0016	.0016	.0015	.0015	.0014	.0014
3.0	.0013	.0013	.0013	.0012	.0012	.0011	.0011	.0011	.0010	.0010
3.1	.0010	.0009	.0009	.0009	.0008	.0008	.0008	.0008	.0007	.0007
3.2	.0007									
3.3	.0005									
3.4	.0003									
3.5	.00023									
3.6	.00016									
3.7	.00011									
3.8	.00007									
3.9	.00005									
4.0	.00003									

TABLE B. TABLE OF CRITICAL VALUES OF t^*

	Level of significance for one-tailed test					
	.10	.05	.025	.01	.005	.0005
df	Level of significance for two-tailed test					
	.20	.10	.05	.02	.01	.001
1	3.078	6.314	12.706	31.821	63.657	636.619
2	1.886	2.920	4.303	6.965	9.925	31.598
3	1.638	2.353	3.182	4.541	5.841	12.941
4	1.533	2.132	2.776	3.747	4.604	8.610
5	1.476	2.015	2.571	3.365	4.032	6.859
6	1.440	1.943	2.447	3.143	3.707	5.959
7	1.415	1.895	2.365	2.998	3.499	5.405
8	1.397	1.860	2.306	2.896	3.355	5.041
9	1.383	1.833	2.262	2.821	3.250	4.781
10	1.372	1.812	2.228	2.764	3.169	4.587
11	1.363	1.796	2.201	2.718	3.106	4.437
12	1.356	1.782	2.179	2.681	3.055	4.318
13	1.350	1.771	2.160	2.650	3.012	4.221
14	1.345	1.761	2.145	2.624	2.977	4.140
15	1.341	1.753	2.131	2.602	2.947	4.073
16	1.337	1.746	2.120	2.583	2.921	4.015
17	1.333	1.740	2.110	2.567	2.898	3.965
18	1.330	1.734	2.101	2.552	2.878	3.922
19	1.328	1.729	2.093	2.539	2.861	3.883
20	1.325	1.725	2.086	2.528	2.845	3.850
21	1.323	1.721	2.080	2.518	2.831	3.819
22	1.321	1.717	2.074	2.508	2.819	3.792
23	1.319	1.714	2.069	2.500	2.807	3.767
24	1.318	1.711	2.064	2.492	2.797	3.745
25	1.316	1.708	2.060	2.485	2.787	3.725
26	1.315	1.706	2.056	2.479	2.779	3.707
27	1.314	1.703	2.052	2.473	2.771	3.690
28	1.313	1.701	2.048	2.467	2.763	3.674
29	1.311	1.699	2.045	2.462	2.756	3.659
30	1.310	1.697	2.042	2.457	2.750	3.646
40	1.303	1.684	2.021	2.423	2.704	3.551
60	1.296	1.671	2.000	2.390	2.660	3.460
120	1.289	1.658	1.980	2.358	2.617	3.373
∞	1.282	1.645	1.960	2.326	2.576	3.291

* Table B is abridged from Table III of Fisher and Yates: *Statistical tables for biological, agricultural, and medical research,* published by Oliver and Boyd Ltd., Edinburgh, by permission of the authors and publishers.

Table C. Table of Critical Values of Chi Square*

df					Probability under H_0 that $\chi^2 \geq$ chi square									
	.99	.98	.95	.90	.80	.70	.50	.30	.20	.10	.05	.02	.01	.001
1	.00016	.00063	.0039	.016	.064	.15	.46	1.07	1.64	2.71	3.84	5.41	6.64	10.83
2	.02	.04	.10	.21	.45	.71	1.39	2.41	3.22	4.60	5.99	7.82	9.21	13.82
3	.12	.18	.35	.58	1.00	1.42	2.37	3.66	4.64	6.25	7.82	9.84	11.34	16.27
4	.30	.43	.71	1.06	1.65	2.20	3.36	4.88	5.99	7.78	9.49	11.67	13.28	18.46
5	.55	.75	1.14	1.61	2.34	3.00	4.35	6.06	7.29	9.24	11.07	13.39	15.09	20.52
6	.87	1.13	1.64	2.20	3.07	3.83	5.35	7.23	8.56	10.64	12.59	15.03	16.81	22.46
7	1.24	1.56	2.17	2.83	3.82	4.67	6.35	8.38	9.80	12.02	14.07	16.62	18.48	24.32
8	1.65	2.03	2.73	3.49	4.59	5.53	7.34	9.52	11.03	13.36	15.51	18.17	20.09	26.12
9	2.09	2.53	3.32	4.17	5.38	6.39	8.34	10.66	12.24	14.68	16.92	19.68	21.67	27.88
10	2.56	3.06	3.94	4.86	6.18	7.27	9.34	11.78	13.44	15.99	18.31	21.16	23.21	29.59
11	3.05	3.61	4.58	5.58	6.99	8.15	10.34	12.90	14.63	17.28	19.68	22.62	24.72	31.26
12	3.57	4.18	5.23	6.30	7.81	9.03	11.34	14.01	15.81	18.55	21.03	24.05	26.22	32.91
13	4.11	4.76	5.89	7.04	8.63	9.93	12.34	15.12	16.98	19.81	22.36	25.47	27.69	34.53
14	4.66	5.37	6.57	7.79	9.47	10.82	13.34	16.22	18.15	21.06	23.68	26.87	29.14	36.12
15	5.23	5.98	7.26	8.55	10.31	11.72	14.34	17.32	19.31	22.31	25.00	28.26	30.58	37.70
16	5.81	6.61	7.96	9.31	11.15	12.62	15.34	18.42	20.46	23.54	26.30	29.63	32.00	39.29
17	6.41	7.26	8.67	10.08	12.00	13.53	16.34	19.51	21.62	24.77	27.59	31.00	33.41	40.75
18	7.02	7.91	9.39	10.86	12.86	14.44	17.34	20.60	22.76	25.99	28.87	32.35	34.80	42.31
19	7.63	8.57	10.12	11.65	13.72	15.35	18.34	21.69	23.90	27.20	30.14	33.69	36.19	43.82
20	8.26	9.24	10.85	12.44	14.58	16.27	19.34	22.78	25.04	28.41	31.41	35.02	37.57	45.32
21	8.90	9.92	11.59	13.24	15.44	17.18	20.34	23.86	26.17	29.62	32.67	36.34	38.93	46.80
22	9.54	10.60	12.34	14.04	16.31	18.10	21.24	24.94	27.30	30.81	33.92	37.66	40.29	48.27
23	10.20	11.29	13.09	14.85	17.19	19.02	22.34	26.02	28.43	32.01	35.17	38.97	41.64	49.73
24	10.86	11.99	13.85	15.66	18.06	19.94	23.34	27.10	29.55	33.20	36.42	40.27	42.98	51.18
25	11.52	12.70	14.61	16.47	18.94	20.87	24.34	28.17	30.68	34.38	37.65	41.57	44.31	52.62
26	12.20	13.41	15.38	17.29	19.82	21.79	25.34	29.25	31.80	35.56	38.88	42.86	45.64	54.05
27	12.88	14.12	16.15	18.11	20.70	22.72	26.34	30.32	32.91	36.74	40.11	44.14	46.96	55.48
28	13.56	14.85	16.93	18.94	21.59	23.65	27.34	31.39	34.03	37.92	41.34	45.42	48.28	56.89
29	14.26	15.57	17.71	19.77	22.48	24.58	28.34	32.46	35.14	39.09	42.56	46.69	49.59	58.30
30	14.95	16.31	18.49	20.60	23.36	25.51	29.34	33.53	36.25	40.26	43.77	47.96	50.89	59.70

* Table C is abridged from Table IV of Fisher and Yates: *Statistical tables for biological, agricultural, and medical research*, published by Oliver and Boyd Ltd., Edinburgh, by permission of the authors and publishers.

TABLE D. TABLE OF PROBABILITIES ASSOCIATED WITH VALUES AS SMALL AS
OBSERVED VALUES OF x IN THE BINOMIAL TEST*

Given in the body of this table are one-tailed probabilities under H_0 for the binomial test when $P = Q = \frac{1}{2}$. To save space, decimal points are omitted in the p's.

N \ x	0	1	2	3	4	5	6	7	8	9	10	11	12	13	14	15
5	031	188	500	812	969	†										
6	016	109	344	656	891	984	†									
7	008	062	227	500	773	938	992	†								
8	004	035	145	363	637	855	965	996	†							
9	002	020	090	254	500	746	910	980	998	†						
10	001	011	055	172	377	623	828	945	989	999	†					
11		006	033	113	274	500	726	887	967	994	†	†				
12		003	019	073	194	387	613	806	927	981	997	†	†			
13		002	011	046	133	291	500	709	867	954	989	998	†	†		
14		001	006	029	090	212	395	605	788	910	971	994	999	†	†		
15			004	018	059	151	304	500	696	849	941	982	996	†	†	†
16			002	011	038	105	227	402	598	773	895	962	989	998	†	†
17			001	006	025	072	166	315	500	685	834	928	975	994	999	†
18			001	004	015	048	119	240	407	593	760	881	952	985	996	999
19				002	010	032	084	180	324	500	676	820	916	968	990	998
20				001	006	021	058	132	252	412	588	748	868	942	979	994
21				001	004	013	039	095	192	332	500	668	808	905	961	987
22					002	008	026	067	143	262	416	584	738	857	933	974
23					001	005	017	047	105	202	339	500	661	798	895	953
24					001	003	011	032	076	154	271	419	581	729	846	924
25						002	007	022	054	115	212	345	500	655	788	885

* Adapted from Table IV, B, of Walker, Helen, and Lev, J. 1953. *Statistical inference.* New York: Holt, p. 458, with the kind permission of the authors and publisher.

† 1.0 or approximately 1.0.

TABLE E. TABLE OF CRITICAL VALUES OF D IN THE KOLMOGOROV-SMIRNOV ONE-SAMPLE TEST*

Sample size (N)	Level of significance for D = maximum $|F_0(X) - S_N(X)|$				
	.20	.15	.10	.05	.01
1	.900	.925	.950	.975	.995
2	.684	.726	.776	.842	.929
3	.565	.597	.642	.708	.828
4	.494	.525	.564	.624	.733
5	.446	.474	.510	.565	.669
6	.410	.436	.470	.521	.618
7	.381	.405	.438	.486	.577
8	.358	.381	.411	.457	.543
9	.339	.360	.388	.432	.514
10	.322	.342	.368	.410	.490
11	.307	.326	.352	.391	.468
12	.295	.313	.338	.375	.450
13	.284	.302	.325	.361	.433
14	.274	.292	.314	.349	.418
15	.266	.283	.304	.338	.404
16	.258	.274	.295	.328	.392
17	.250	.266	.286	.318	.381
18	.244	.259	.278	.309	.371
19	.237	.252	.272	.301	.363
20	.231	.246	.264	.294	.356
25	.21	.22	.24	.27	.32
30	.19	.20	.22	.24	.29
35	.18	.19	.21	.23	.27
Over 35	$\dfrac{1.07}{\sqrt{N}}$	$\dfrac{1.14}{\sqrt{N}}$	$\dfrac{1.22}{\sqrt{N}}$	$\dfrac{1.36}{\sqrt{N}}$	$\dfrac{1.63}{\sqrt{N}}$

* Adapted from Massey, F. J., Jr. 1951. The Kolmogorov-Smirnov test for goodness of fit. *J. Amer. Statist. Ass.*, **46**, 70, with the kind permission of the author and publisher.

TABLE F. TABLE OF CRITICAL VALUES OF r IN THE RUNS TEST*

Given in the bodies of Table F_I and Table F_{II} are various critical values of r for various values of n_1 and n_2. For the one-sample runs test, any value of r which is equal to or smaller than that shown in Table F_I or equal to or larger than that shown in Table F_{II} is significant at the .05 level. For the Wald-Wolfowitz two-sample runs test, any value of r which is equal to or smaller than that shown in Table F_I is significant at the .05 level.

Table F_I

n_1 \ n_2	2	3	4	5	6	7	8	9	10	11	12	13	14	15	16	17	18	19	20
2											2	2	2	2	2	2	2	2	2
3					2	2	2	2	2	2	2	2	2	3	3	3	3	3	3
4				2	2	2	3	3	3	3	3	3	3	3	4	4	4	4	4
5			2	2	3	3	3	3	3	4	4	4	4	4	4	4	5	5	5
6		2	2	3	3	3	3	4	4	4	4	5	5	5	5	5	5	6	6
7		2	2	3	3	3	4	4	5	5	5	5	5	6	6	6	6	6	6
8		2	3	3	3	4	4	5	5	5	6	6	6	6	6	7	7	7	7
9		2	3	3	4	4	5	5	5	6	6	6	7	7	7	7	8	8	8
10		2	3	3	4	5	5	5	6	6	7	7	7	7	8	8	8	8	9
11		2	3	4	4	5	5	6	6	7	7	7	8	8	8	9	9	9	9
12	2	2	3	4	4	5	6	6	7	7	7	8	8	8	9	9	9	10	10
13	2	2	3	4	5	5	6	6	7	7	8	8	9	9	9	10	10	10	10
14	2	2	3	4	5	5	6	7	7	8	8	9	9	9	10	10	10	11	11
15	2	3	3	4	5	6	6	7	7	8	8	9	9	10	10	11	11	11	12
16	2	3	4	4	5	6	6	7	8	8	9	9	10	10	11	11	11	12	12
17	2	3	4	4	5	6	7	7	8	9	9	10	10	11	11	11	12	12	13
18	2	3	4	5	5	6	7	8	8	9	9	10	10	11	11	12	12	13	13
19	2	3	4	5	6	6	7	8	8	9	10	10	11	11	12	12	13	13	13
20	2	3	4	5	6	6	7	8	9	9	10	10	11	12	12	13	13	13	14

* Adapted from Swed, Frieda S., and Eisenhart, C. 1943. Tables for testing randomness of grouping in a sequence of alternatives. *Ann. Math. Statist.*, **14**, 83–86, with the kind permission of the authors and publisher.

TABLE F. TABLE OF CRITICAL VALUES OF r IN THE RUNS TEST* (*Continued*)

Table F_{II}

n_1 \ n_2	2	3	4	5	6	7	8	9	10	11	12	13	14	15	16	17	18	19	20
2																			
3																			
4			9	9															
5		9	10	10	11	11													
6		9	10	11	12	12	13	13	13	13									
7				11	12	13	13	14	14	14	14	15	15	15					
8				11	12	13	14	14	15	15	16	16	16	16	17	17	17	17	17
9					13	14	14	15	16	16	16	17	17	18	18	18	18	18	18
10					13	14	15	16	16	17	17	18	18	18	19	19	19	20	20
11					13	14	15	16	17	17	18	19	19	19	20	20	20	21	21
12					13	14	16	16	17	18	19	19	20	20	21	21	21	22	22
13						15	16	17	18	19	19	20	20	21	21	22	22	23	23
14						15	16	17	18	19	20	20	21	22	22	23	23	23	24
15						15	16	18	18	19	20	21	22	22	23	23	23	24	25
16							17	18	19	20	21	21	22	23	23	24	25	25	25
17							17	18	19	20	21	22	23	23	24	25	25	26	26
18							17	18	19	20	21	22	23	24	25	25	26	26	27
19							17	18	20	21	22	23	23	24	25	26	26	27	27
20							17	18	20	21	22	23	24	25	25	26	27	27	28

* Adapted from Swed, Frieda S., and Eisenhart, C. 1943. Tables for testing randomness of grouping in a sequence of alternatives. *Ann. Math. Statist.*, **14**, 83–86, with the kind permission of the authors and publisher.

TABLE G. TABLE OF CRITICAL VALUES OF T IN THE WILCOXON
MATCHED-PAIRS SIGNED-RANKS TEST*

	Level of significance for one-tailed test		
	.025	.01	.005
N	Level of significance for two-tailed test		
	.05	.02	.01
6	0	—	—
7	2	0	—
8	4	2	0
9	6	3	2
10	8	5	3
11	11	7	5
12	14	10	7
13	17	13	10
14	21	16	13
15	25	20	16
16	30	24	20
17	35	28	23
18	40	33	28
19	46	38	32
20	52	43	38
21	59	49	43
22	66	56	49
23	73	62	55
24	81	69	61
25	89	77	68

* Adapted from Table I of Wilcoxon, F. 1949. *Some rapid approximate statistical procedures.* New York: American Cyanamid Company, p. 13, with the kind permission of the author and publisher.

TABLE H. TABLE OF CRITICAL VALUES FOR THE WALSH TEST[*]

N	Significance level of tests		Tests	
			Two-tailed: accept $\mu_1 \neq 0$ if either	
	One-tailed	Two-tailed	One-tailed: accept $\mu_1 < 0$ if	One-tailed: accept $\mu_1 > 0$ if
4	.062	.125	$d_4 < 0$	$d_1 > 0$
5	.062	.125	$\frac{1}{2}(d_4 + d_5) < 0$	$\frac{1}{2}(d_1 + d_2) > 0$
	.031	.062	$d_5 < 0$	$d_1 > 0$
6	.047	.094	max $[d_5, \frac{1}{2}(d_4 + d_6)] < 0$	min $[d_2, \frac{1}{2}(d_1 + d_3)] > 0$
	.031	.062	$\frac{1}{2}(d_5 + d_6) < 0$	$\frac{1}{2}(d_1 + d_2) > 0$
	.016	.031	$d_6 < 0$	$d_1 > 0$
7	.055	.109	max $[d_5, \frac{1}{2}(d_4 + d_7)] < 0$	min $[d_3, \frac{1}{2}(d_1 + d_4)] > 0$
	.023	.047	max $[d_6, \frac{1}{2}(d_5 + d_7)] < 0$	min $[d_2, \frac{1}{2}(d_1 + d_3)] > 0$
	.016	.031	$\frac{1}{2}(d_6 + d_7) < 0$	$\frac{1}{2}(d_1 + d_2) > 0$
	.008	.016	$d_7 < 0$	$d_1 > 0$
8	.043	.086	max $[d_6, \frac{1}{2}(d_4 + d_8)] > 0$	min $[d_3, \frac{1}{2}(d_1 + d_5)] > 0$
	.027	.055	max $[d_6, \frac{1}{2}(d_5 + d_8)] < 0$	min $[d_3, \frac{1}{2}(d_1 + d_4)] > 0$
	.012	.023	max $[d_7, \frac{1}{2}(d_6 + d_8)] < 0$	min $[d_2, \frac{1}{2}(d_1 + d_3)] > 0$
	.008	.016	$\frac{1}{2}(d_7 + d_8) < 0$	$\frac{1}{2}(d_1 + d_2) > 0$
	.004	.008	$d_8 < 0$	$d_1 > 0$
9	.051	.102	max $[d_6, \frac{1}{2}(d_4 + d_9)] < 0$	min $[d_4, \frac{1}{2}(d_1 + d_6)] > 0$
	.022	.043	max $[d_7, \frac{1}{2}(d_5 + d_9)] < 0$	min $[d_3, \frac{1}{2}(d_1 + d_5)] > 0$
	.010	.020	max $[d_8, \frac{1}{2}(d_6 + d_9)] < 0$	min $[d_2, \frac{1}{2}(d_1 + d_5)] > 0$
	.006	.012	max $[d_8, \frac{1}{2}(d_7 + d_9)] < 0$	min $[d_2, \frac{1}{2}(d_1 + d_3)] > 0$
	.004	.008	$\frac{1}{2}(d_8 + d_9) < 0$	$\frac{1}{2}(d_1 + d_2) > 0$
10	.056	.111	max $[d_6, \frac{1}{2}(d_4 + d_{10})] < 0$	min $[d_5, \frac{1}{2}(d_1 + d_7)] > 0$
	.025	.051	max $[d_7, \frac{1}{2}(d_5 + d_{10})] < 0$	min $[d_4, \frac{1}{2}(d_1 + d_6)] > 0$
	.011	.021	max $[d_8, \frac{1}{2}(d_6 + d_{10})] < 0$	min $[d_3, \frac{1}{2}(d_1 + d_5)] > 0$
	.005	.010	max $[d_9, \frac{1}{2}(d_6 + d_{10})] < 0$	min $[d_2, \frac{1}{2}(d_1 + d_5)] > 0$
11	.048	.097	max $[d_7, \frac{1}{2}(d_4 + d_{11})] < 0$	min $[d_5, \frac{1}{2}(d_1 + d_8)] > 0$
	.028	.056	max $[d_7, \frac{1}{2}(d_5 + d_{11})] < 0$	min $[d_5, \frac{1}{2}(d_1 + d_7)] > 0$
	.011	.021	max $[\frac{1}{2}(d_6 + d_{11}), \frac{1}{2}(d_8 + d_9)] < 0$	min $[\frac{1}{2}(d_1 + d_6), \frac{1}{2}(d_3 + d_4)] > 0$
	.005	.011	max $[d_9, \frac{1}{2}(d_7 + d_{11})] < 0$	min $[d_3, \frac{1}{2}(d_1 + d_5)] > 0$
12	.047	.094	max $[\frac{1}{2}(d_4 + d_{12}), \frac{1}{2}(d_5 + d_{11})] < 0$	min $[\frac{1}{2}(d_1 + d_9), \frac{1}{2}(d_2 + d_5)] > 0$
	.024	.048	max $[d_8, \frac{1}{2}(d_5 + d_{12})] < 0$	min $[d_5, \frac{1}{2}(d_1 + d_8)] > 0$
	.010	.020	max $[d_9, \frac{1}{2}(d_6 + d_{12})] < 0$	min $[d_4, \frac{1}{2}(d_1 + d_7)] > 0$
	.005	.011	max $[\frac{1}{2}(d_7 + d_{12}), \frac{1}{2}(d_9 + d_{10})] < 0$	min $[\frac{1}{2}(d_1 + d_6), \frac{1}{2}(d_3 + d_4)] > 0$
13	.047	.094	max $[\frac{1}{2}(d_4 + d_{13}), \frac{1}{2}(d_5 + d_{12})] < 0$	min $[\frac{1}{2}(d_1 + d_{10}), \frac{1}{2}(d_2 + d_9)] > 0$
	.023	.047	max $[\frac{1}{2}(d_5 + d_{13}), \frac{1}{2}(d_6 + d_{12})] < 0$	min $[\frac{1}{2}(d_1 + d_8), \frac{1}{2}(d_2 + d_5)] > 0$
	.010	.020	max $[\frac{1}{2}(d_6 + d_{13}), \frac{1}{2}(d_9 + d_{10})] < 0$	min $[\frac{1}{2}(d_1 + d_5), \frac{1}{2}(d_4 + d_5)] > 0$
	.005	.010	max $[d_{10}, \frac{1}{2}(d_7 + d_{13})] < 0$	min $[d_4, \frac{1}{2}(d_1 + d_7)] > 0$
14	.047	.094	max $[\frac{1}{2}(d_4 + d_{14}), \frac{1}{2}(d_5 + d_{13})] < 0$	min $[\frac{1}{2}(d_1 + d_{11}), \frac{1}{2}(d_2 + d_{10})] > 0$
	.023	.047	max $[\frac{1}{2}(d_5 + d_{14}), \frac{1}{2}(d_6 + d_{13})] < 0$	min $[\frac{1}{2}(d_1 + d_{10}), \frac{1}{2}(d_2 + d_9)] > 0$
	.010	.020	max $[d_{10}, \frac{1}{2}(d_6 + d_{14})] < 0$	min $[d_5, \frac{1}{2}(d_1 + d_9)] > 0$
	.005	.010	max $[\frac{1}{2}(d_7 + d_{14}), \frac{1}{2}(d_{10} + d_{11})] < 0$	min $[\frac{1}{2}(d_1 + d_8), \frac{1}{2}(d_4 + d_5)] > 0$
15	.047	.094	max $[\frac{1}{2}(d_4 + d_{15}), \frac{1}{2}(d_5 + d_{14})] < 0$	min $[\frac{1}{2}(d_1 + d_{12}), \frac{1}{2}(d_2 + d_{11})] > 0$
	.023	.047	max $[\frac{1}{2}(d_5 + d_{15}), \frac{1}{2}(d_6 + d_{14})] < 0$	min $[\frac{1}{2}(d_1 + d_{11}), \frac{1}{2}(d_2 + d_{10})] > 0$
	.010	.020	max $[\frac{1}{2}(d_6 + d_{15}), \frac{1}{2}(d_{10} + d_{11})] < 0$	min $[\frac{1}{2}(d_1 + d_{10}), \frac{1}{2}(d_5 + d_6)] > 0$
	.005	.010	max $[d_{11}, \frac{1}{2}(d_7 + d_{15})] < 0$	min $[d_5, \frac{1}{2}(d_1 + d_9)] > 0$

[*] Adapted from Walsh, J. E. 1949. Applications of some significance tests for the median which are valid under very general conditions. *J. Amer. Statist. Ass.*, **44**, 343, with the kind permission of the author and the publisher.

TABLE I. TABLE OF CRITICAL VALUES OF D (OR C) IN THE
FISHER TEST*,†

Totals in right margin		B (or A)†	Level of significance			
			.05	.025	.01	.005
$A + B = 3$	$C + D = 3$	3	0	—	—	—
$A + B = 4$	$C + D = 4$	4	0	0	—	—
	$C + D = 3$	4	0	—	—	—
$A + B = 5$	$C + D = 5$	5	1	1	0	0
		4	0	0	—	—
	$C + D = 4$	5	1	0	0	—
		4	0	—	—	—
	$C + D = 3$	5	0	0	—	—
	$C + D = 2$	5	0	—	—	—
$A + B = 6$	$C + D = 6$	6	2	1	1	0
		5	1	0	0	—
		4	0	—	—	—
	$C + D = 5$	6	1	0	0	0
		5	0	0	—	—
		4	0	—	—	—
	$C + D = 4$	6	1	0	0	0
		5	0	0	—	—
	$C + D = 3$	6	0	0	—	—
		5	0	—	—	—
	$C + D = 2$	6	0	—	—	—
$A + B = 7$	$C + D = 7$	7	3	2	1	1
		6	1	1	0	0
		5	0	0	—	—
		4	0	—	—	—
	$C + D = 6$	7	2	2	1	1
		6	1	0	0	0
		5	0	0	—	—
		4	0	—	—	—
	$C + D = 5$	7	2	1	0	0
		6	1	0	0	—
		5	0	—	—	—
	$C + D = 4$	7	1	1	0	0
		6	0	0	—	—
		5	0	—	—	—
	$C + D = 3$	7	0	0	0	—
		6	0	—	—	—
	$C + D = 2$	7	0	—	—	—

* Adapted from Finney, D. J. 1948. The Fisher-Yates test of significance in 2×2 contingency tables. *Biometrika*, **35**, 149–154, with the kind permission of the author and the publisher.

TABLE I. TABLE OF CRITICAL VALUES OF D (OR C) IN THE
FISHER TEST*,† (*Continued*)

Totals in right margin		B (or A)†	Level of significance			
			.05	.025	.01	.005
$A + B = 8$	$C + D = 8$	8	4	3	2	2
		7	2	2	1	0
		6	1	1	0	0
		5	0	0	—	—
		4	0	—	—	—
	$C + D = 7$	8	3	2	2	1
		7	2	1	1	0
		6	1	0	0	—
		5	0	0	—	—
	$C + D = 6$	8	2	2	1	1
		7	1	1	0	0
		6	0	0	0	—
		5	0	—	—	—
	$C + D = 5$	8	2	1	1	0
		7	1	0	0	0
		6	0	0	—	—
		5	0	—	—	—
	$C + D = 4$	8	1	1	0	0
		7	0	0	—	—
		6	0	—	—	—
	$C + D = 3$	8	0	0	0	—
		7	0	0	—	—
	$C + D = 2$	8	0	0	—	—
$A + B = 9$	$C + D = 9$	9	5	4	3	3
		8	3	3	2	1
		7	2	1	1	0
		6	1	1	0	0
		5	0	0	—	—
		4	0	—	—	—
	$C + D = 8$	9	4	3	3	2
		8	3	2	1	1
		7	2	1	0	0
		6	1	0	0	—
		5	0	0	—	—
	$C + D = 7$	9	3	3	2	2
		8	2	2	1	0
		7	1	1	0	0
		6	0	0	—	—
		5	0	—	—	—

† When B is entered in the middle column, the significance levels are for D. When A is used in place of B, the significance levels are for C.

TABLE I. TABLE OF CRITICAL VALUES OF D (OR C) IN THE
FISHER TEST*,† (*Continued*)

Totals in right margin		B (or A)†	Level of significance			
			.05	.025	.01	.005
$A + B = 9$	$C + D = 6$	9	3	2	1	1
		8	2	1	0	0
		7	1	0	0	—
		6	0	0	—	—
		5	0	—	—	—
	$C + D = 5$	9	2	1	1	1
		8	1	1	0	0
		7	0	0	—	—
		6	0	—	—	—
	$C + D = 4$	9	1	1	0	0
		8	0	0	0	—
		7	0	0	—	—
		6	0	—	—	—
	$C + D = 3$	9	1	0	0	0
		8	0	0	—	—
		7	0	—	—	—
	$C + D = 2$	9	0	0	—	—
$A + B = 10$	$C + D = 10$	10	6	5	4	3
		9	4	3	3	2
		8	3	2	1	1
		7	2	1	1	0
		6	1	0	0	—
		5	0	0	—	—
		4	0	—	—	—
	$C + D = 9$	10	5	4	3	3
		9	4	3	2	2
		8	2	2	1	1
		7	1	1	0	0
		6	1	0	0	—
		5	0	0	—	—
	$C + D = 8$	10	4	4	3	2
		9	3	2	2	1
		8	2	1	1	0
		7	1	1	0	0
		6	0	0	—	—
		5	0	—	—	—
	$C + D = 7$	10	3	3	2	2
		9	2	2	1	1
		8	1	1	0	0
		7	1	0	0	—
		6	0	0	—	—
		5	0	—	—	—

* Adapted from Finney, D. J. 1948. The Fisher-Yates test of significance in
2×2 contingency tables. *Biometrika*, **35**, 149–154, with the kind permission of the
author and the publisher.

TABLE I. TABLE OF CRITICAL VALUES OF D (OR C) IN THE
FISHER TEST*,† (*Continued*)

Totals in right margin		B (or A)†	Level of significance			
			.05	.025	.01	.005
$A + B = 10$	$C + D = 6$	10	3	2	2	1
		9	2	1	1	0
		8	1	1	0	0
		7	0	0	—	—
		6	0	—	—	—
	$C + D = 5$	10	2	2	1	1
		9	1	1	0	0
		8	1	0	0	—
		7	0	0	—	—
		6	0	—	—	—
	$C + D = 4$	10	1	1	0	0
		9	1	0	0	0
		8	0	0	—	—
		7	0	—	—	—
	$C + D = 3$	10	1	0	0	0
		9	0	0	—	—
		8	0	—	—	—
	$C + D = 2$	10	0	0	—	—
		9	0	—	—	—
$A + B = 11$	$C + D = 11$	11	7	6	5	4
		10	5	4	3	3
		9	4	3	2	2
		8	3	2	1	1
		7	2	1	0	0
		6	1	0	0	—
		5	0	0	—	—
		4	0	—	—	—
	$C + D = 10$	11	6	5	4	4
		10	4	4	3	2
		9	3	3	2	1
		8	2	2	1	0
		7	1	1	0	0
		6	1	0	0	—
		5	0	—	—	—
	$C + D = 9$	11	5	4	4	3
		10	4	3	2	2
		9	3	2	1	1
		8	2	1	1	0
		7	1	1	0	0
		6	0	0	—	—
		5	0	—	—	—

† When B is entered in the middle column, the significance levels are for D. When A is used in place of B, the significance levels are for C.

TABLE I. TABLE OF CRITICAL VALUES OF D (OR C) IN THE FISHER TEST*,† (*Continued*)

Totals in right margin		B (or A)†	Level of significance			
			.05	.025	.01	.005
$A + B = 11$	$C + D = 8$	11	4	4	3	3
		10	3	3	2	1
		9	2	2	1	1
		8	1	1	0	0
		7	1	0	0	—
		6	0	0	—	—
		5	0	—	—	—
	$C + D = 7$	11	4	3	2	2
		10	3	2	1	1
		9	2	1	1	0
		8	1	1	0	0
		7	0	0	—	—
		6	0	0	—	—
	$C + D = 6$	11	3	2	2	1
		10	2	1	1	0
		9	1	1	0	0
		8	1	0	0	—
		7	0	0	—	—
		6	0	—	—	—
	$C + D = 5$	11	2	2	1	1
		10	1	1	0	0
		9	1	0	0	0
		8	0	0	—	—
		7	0	—	—	—
	$C + D = 4$	11	1	1	1	0
		10	1	0	0	0
		9	0	0	—	—
		8	0	—	—	—
	$C + D = 3$	11	1	0	0	0
		10	0	0	—	—
		9	0	—	—	—
	$C + D = 2$	11	0	0	—	—
		10	0	—	—	—
$A + B = 12$	$C + D = 12$	12	8	7	6	5
		11	6	5	4	4
		10	5	4	3	2
		9	4	3	2	1
		8	3	2	1	1
		7	2	1	0	0
		6	1	0	0	—
		5	0	0	—	—
		4	0	—	—	—

* Adapted from Finney, D. J. 1948. The Fisher-Yates test of significance in 2×2 contingency tables. *Biometrika*, **35**, 149–154, with the kind permission of the author and the publisher.

TABLE I. TABLE OF CRITICAL VALUES OF D (OR C) IN THE
FISHER TEST*,† (*Continued*)

Totals in right margin		B (or A)†	Level of significance			
			.05	.025	.01	.005
$A + B = 12$	$C + D = 11$	12	7	6	5	5
		11	5	5	4	3
		10	4	3	2	2
		9	3	2	2	1
		8	2	1	1	0
		7	1	1	0	0
		6	1	0	0	—
		5	0	0	—	—
	$C + D = 10$	12	6	5	5	4
		11	5	4	3	3
		10	4	3	2	2
		9	3	2	1	1
		8	2	1	0	0
		7	1	0	0	0
		6	0	0	—	—
		5	0	—	—	—
	$C + D = 9$	12	5	5	4	3
		11	4	3	3	2
		10	3	2	2	1
		9	2	2	1	0
		8	1	1	0	0
		7	1	0	0	—
		6	0	0	—	—
		5	0	—	—	—
	$C + D = 8$	12	5	4	3	3
		11	3	3	2	2
		10	2	2	1	1
		9	2	1	1	0
		8	1	1	0	0
		7	0	0	—	—
		6	0	0	—	—
	$C + D = 7$	12	4	3	3	2
		11	3	2	2	1
		10	2	1	1	0
		9	1	1	0	0
		8	1	0	0	—
		7	0	0	—	—
		6	0	—	—	—

† When B is entered in the middle column, the significance levels are for D. When A is used in place of B, the significance levels are for C.

TABLE I. TABLE OF CRITICAL VALUES OF D (OR C) IN THE
FISHER TEST*,† (*Continued*)

Totals in right margin		B (or A)†	Level of significance			
			.05	.025	.01	.005
$A + B = 12$	$C + D = 6$	12	3	3	2	2
		11	2	2	1	1
		10	1	1	0	0
		9	1	0	0	0
		8	0	0	—	—
		7	0	0	—	—
		6	0	—	—	—
	$C + D = 5$	12	2	2	1	1
		11	1	1	1	0
		10	1	0	0	0
		9	0	0	0	—
		8	0	0	—	—
		7	0	—	—	—
	$C + D = 4$	12	2	1	1	0
		11	1	0	0	0
		10	0	0	0	—
		9	0	0	—	—
		8	0	—	—	—
	$C + D = 3$	12	1	0	0	0
		11	0	0	0	—
		10	0	0	—	—
		9	0	—	—	—
	$C + D = 2$	12	0	0	—	—
		11	0	—	—	—
$A + B = 13$	$C + D = 13$	13	9	8	7	6
		12	7	6	5	4
		11	6	5	4	3
		10	4	4	3	2
		9	3	3	2	1
		8	2	2	1	0
		7	2	1	0	0
		6	1	0	0	—
		5	0	0	—	—
		4	0	—	—	—
	$C + D = 12$	13	8	7	6	5
		12	6	5	5	4
		11	5	4	3	3
		10	4	3	2	2
		9	3	2	1	1
		8	2	1	1	0
		7	1	1	0	0
		6	1	0	0	—
		5	0	0	—	—

* Adapted from Finney, D. J. 1948. The Fisher-Yates test of significance in 2×2 contingency tables. *Biometrika*, **35**, 149–154, with the kind permission of the author and the publisher.

TABLE I. TABLE OF CRITICAL VALUES OF D (OR C) IN THE
FISHER TEST*,† (*Continued*)

Totals in right margin		B (or A)†	Level of significance			
			.05	.025	.01	.005
$A + B = 13$	$C + D = 11$	13	7	6	5	5
		12	6	5	4	3
		11	4	4	3	2
		10	3	3	2	1
		9	3	2	1	1
		8	2	1	0	0
		7	1	0	0	0
		6	0	0	—	—
		5	0	—	—	—
	$C + D = 10$	13	6	6	5	4
		12	5	4	3	3
		11	4	3	2	2
		10	3	2	1	1
		9	2	1	1	0
		8	1	1	0	0
		7	1	0	0	—
		6	0	0	—	—
		5	0	—	—	—
	$C + D = 9$	13	5	5	4	4
		12	4	4	3	2
		11	3	3	2	1
		10	2	2	1	1
		9	2	1	0	0
		8	1	1	0	0
		7	0	0	—	—
		6	0	0	—	—
		5	0	—	—	—
	$C + D = 8$	13	5	4	3	3
		12	4	3	2	2
		11	3	2	1	1
		10	2	1	1	0
		9	1	1	0	0
		8	1	0	0	—
		7	0	0	—	—
		6	0	—	—	—
	$C + D = 7$	13	4	3	3	2
		12	3	2	2	1
		11	2	2	1	1
		10	1	1	0	0
		9	1	0	0	0
		8	0	0	—	—
		7	0	0	—	—
		6	0	—	—	—

† When B is entered in the middle column, the significance levels are for D. When A is used in place of B, the significance levels are for C.

TABLE I. TABLE OF CRITICAL VALUES OF D (OR θ) IN THE
FISHER TEST*,† (Continued)

Totals in right margin		B (or A)†	Level of significance			
			.05	.025	.01	.005
$A + B = 13$	$C + D = 6$	13	3	3	2	2
		12	2	2	1	1
		11	2	1	1	0
		10	1	1	0	0
		9	1	0	0	—
		8	0	0	—	—
		7	0	—	—	—
	$C + D = 5$	13	2	2	1	1
		12	2	1	1	0
		11	1	1	0	0
		10	1	0	0	—
		9	0	0	—	—
		8	0	—	—	—
	$C + D = 4$	13	2	1	1	0
		12	1	1	0	0
		11	0	0	0	—
		10	0	0	—	—
		9	0	—	—	—
	$C + D = 3$	13	1	1	0	0
		12	0	0	0	—
		11	0	0	—	—
		10	0	—	—	—
	$C + D = 2$	13	0	0	0	—
		12	0	—	—	—
$A + B = 14$	$C + D = 14$	14	10	9	8	7
		13	8	7	6	5
		12	6	6	5	4
		11	5	4	3	3
		10	4	3	2	2
		9	3	2	2	1
		8	2	2	1	0
		7	1	1	0	0
		6	1	0	0	—
		5	0	0	—	—
		4	0	—	—	—

* Adapted from Finney, D. J. 1948. The Fisher-Yates test of significance in 2 × 2 contingency tables. *Biometrika*, **35**, 149–154, with the kind permission of the author and the publisher.

TABLE I. TABLE OF CRITICAL VALUES OF D (OR C) IN THE
FISHER TEST*,† (*Continued*)

Totals in right margin		B (or A)†	Level of significance			
			.05	.025	.01	.005
$A + B = 14$	$C + D = 13$	14	9	8	7	6
		13	7	6	5	5
		12	6	5	4	3
		11	5	4	3	2
		10	4	3	2	2
		9	3	2	1	1
		8	2	1	1	0
		7	1	1	0	0
		6	1	0	—	—
		5	0	0	—	—
	$C + D = 12$	14	8	7	6	6
		13	6	6	5	4
		12	5	4	4	3
		11	4	3	3	2
		10	3	3	2	1
		9	2	2	1	1
		8	2	1	0	0
		7	1	0	0	—
		6	0	0	—	—
		5	0	—	—	—
	$C + D = 11$	14	7	6	6	5
		13	6	5	4	4
		12	5	4	3	3
		11	4	3	2	2
		10	3	2	1	1
		9	2	1	1	0
		8	1	1	0	0
		7	1	0	0	—
		6	0	0	—	—
		5	0	—	—	—
	$C + D = 10$	14	6	6	5	4
		13	5	4	4	3
		12	4	3	3	2
		11	3	3	2	1
		10	2	2	1	1
		9	2	1	0	0
		8	1	1	0	0
		7	0	0	0	—
		6	0	0	—	—
		5	0	—	—	—

† When B is entered in the middle column, the significance levels are for D. When A is used in place of B, the significance levels yre for C.

TABLE I. TABLE OF CRITICAL VALUES OF *D* (OR *C*) IN THE
FISHER TEST*,† (*Continued*)

Totals in right margin		B (or A)†	Level of significance			
			.05	.025	.01	.005
$A + B = 14$	$C + D = 9$	14	6	5	4	4
		13	4	4	3	3
		12	3	3	2	2
		11	3	2	1	1
		10	2	1	1	0
		9	1	1	0	0
		8	1	0	0	—
		7	0	0	—	—
		6	0	—	—	—
	$C + D = 8$	14	5	4	4	3
		13	4	3	2	2
		12	3	2	2	1
		11	2	2	1	1
		10	2	1	0	0
		9	1	0	0	0
		8	0	0	0	—
		7	0	0	—	—
		6	0	—	—	—
	$C + D = 7$	14	4	3	3	2
		13	3	2	2	1
		12	2	2	1	1
		11	2	1	1	0
		10	1	1	0	0
		9	1	0	0	—
		8	0	0	—	—
		7	0	—	—	—
	$C + D = 6$	14	3	3	2	2
		13	2	2	1	1
		12	2	1	1	0
		11	1	1	0	0
		10	1	0	0	—
		9	0	0	—	—
		8	0	0	—	—
		7	0	—	—	—
	$C + D = 5$	14	2	2	1	1
		13	2	1	1	0
		12	1	1	0	0
		11	1	0	0	0
		10	0	0	—	—
		9	0	0	—	—
		8	0	—	—	—

* Adapted from Finney, D. J. 1948. The Fisher-Yates test of significance in
2 × 2 contingency tables. *Biometrika,* **35,** 149–154, with the kind permission of the
author and the publisher.

TABLE I. TABLE OF CRITICAL VALUES OF D (OR C) IN THE
FISHER TEST*,† (*Continued*)

Totals in right margin		B (or A)†	Level of significance			
			.05	.025	.01	.005
$A + B = 14$	$C + D = 4$	14	2	1	1	1
		13	1	1	0	0
		12	1	0	0	0
		11	0	0	—	—
		10	0	0	—	—
		9	0	—	—	—
	$C + D = 3$	14	1	1	0	0
		13	0	0	0	—
		12	0	0	—	—
		11	0	—	—	—
	$C + D = 2$	14	0	0	0	—
		13	0	0	—	—
		12	0	—	—	—
$A + B = 15$	$C + D = 15$	15	11	10	9	8
		14	9	8	7	6
		13	7	6	5	5
		12	6	5	4	4
		11	5	4	3	3
		10	4	3	2	2
		9	3	2	1	1
		8	2	1	1	0
		7	1	1	0	0
		6	1	0	0	—
		5	0	0	—	—
		4	0	—	—	—
	$C + D = 14$	15	10	9	8	7
		14	8	7	6	6
		13	7	6	5	4
		12	6	5	4	3
		11	5	4	3	2
		10	4	3	2	1
		9	3	2	1	1
		8	2	1	1	0
		7	1	1	0	0
		6	1	0	—	—
		5	0	—	—	—

† When B is entered in the middle column, the significance levels are for D. When A is used in place of B, the significance levels are for C.

TABLE I. TABLE OF CRITICAL VALUES OF D (OR C) IN THE
FISHER TEST*,† (Continued)

Totals in right margin		B (or A)†	Level of significance			
			.05	.025	.01	.005
$A + B = 15$	$C + D = 13$	15	9	8	7	7
		14	7	7	6	5
		13	6	5	4	4
		12	5	4	3	3
		11	4	3	2	2
		10	3	2	2	1
		9	2	2	1	0
		8	2	1	0	0
		7	1	0	0	—
		6	0	0	—	—
		5	0	—	—	—
	$C + D = 12$	15	8	7	7	6
		14	7	6	5	4
		13	6	5	4	3
		12	5	4	3	2
		11	4	3	2	2
		10	3	2	1	1
		9	2	1	1	0
		8	1	1	0	0
		7	1	0	0	—
		6	0	0	—	—
		5	0	—	—	—
	$C + D = 11$	15	7	7	6	5
		14	6	5	4	4
		13	5	4	3	3
		12	4	3	2	2
		11	3	2	2	1
		10	2	2	1	1
		9	2	1	0	0
		8	1	1	0	0
		7	1	0	0	—
		6	0	0	—	—
		5	0	—	—	—
	$C + D = 10$	15	6	6	5	5
		14	5	5	4	3
		13	4	4	3	2
		12	3	3	2	2
		11	3	2	1	1
		10	2	1	1	0
		9	1	1	0	0
		8	1	0	0	—
		7	0	0	—	—
		6	0	—	—	—

* Adapted from Finney, D. J. 1948. The Fisher-Yates test of significance in
2 × 2 contingency tables. *Biometrika*, **35**, 149–154, with the kind permission of the
author and the publisher.

TABLE I. TABLE OF CRITICAL VALUES OF D (OR C) IN THE
FISHER TEST*,† (*Continued*)

Totals in right margin		B (or A)†	Level of significance			
			.05	.025	.01	.005
$A + B = 15$	$C + D = 9$	15	6	5	4	4
		14	5	4	3	3
		13	4	3	2	2
		12	3	2	2	1
		11	2	2	1	1
		10	2	1	0	0
		9	1	1	0	0
		8	1	0	0	—
		7	0	0	—	—
		6	0	—	—	—
	$C + D = 8$	15	5	4	4	3
		14	4	3	3	2
		13	3	2	2	1
		12	2	2	1	1
		11	2	1	1	0
		10	1	1	0	0
		9	1	0	0	—
		8	0	0	—	—
		7	0	—	—	—
		6	0	—	—	—
	$C + D = 7$	15	4	4	3	3
		14	3	3	2	2
		13	2	2	1	1
		12	2	1	1	0
		11	1	1	0	0
		10	1	0	0	0
		9	0	0	—	—
		8	0	0	—	—
		7	0	—	—	—
	$C + D = 6$	15	3	3	2	2
		14	2	2	1	1
		13	2	1	1	0
		12	1	1	0	0
		11	1	0	0	0
		10	0	0	0	—
		9	0	0	—	—
		8	0	—	—	—
	$C + D = 5$	15	2	2	2	1
		14	2	1	1	1
		13	1	1	0	0
		12	1	0	0	0
		11	0	0	0	—
		10	0	0	—	—
		9	0	—	—	—

† When B is entered in the middle column, the significance levels are for D. When A is used in place of B, the significance levels are for C.

TABLE I. TABLE OF CRITICAL VALUES OF D (OR C) IN THE
FISHER TEST*,† (*Continued*)

Totals in right margin		B (or A)†	Level of significance			
			.05	.025	.01	.005
$A + B = 15$	$C + D = 4$	15	2	1	1	1
		14	1	1	0	0
		13	1	0	0	0
		12	0	0	0	—
		11	0	0	—	—
		10	0	—	—	—
	$C + D = 3$	15	1	1	0	0
		14	0	0	0	0
		13	0	0	—	—
		12	0	0	—	—
		11	0	—	—	—
	$C + D = 2$	15	0	0	0	—
		14	0	0	—	—
		13	0	—	—	—

* Adapted from Finney, D. J. 1948. The Fisher-Yates test of significance in 2 × 2 contingency tables. *Biometrika*, **35**, 149–154, with the kind permission of the author and the publisher.

† When B is entered in the middle column, the significance levels are for D. When A is used in place of B, the significance levels are for C.

TABLE J. TABLE OF PROBABILITIES ASSOCIATED WITH VALUES AS SMALL AS OBSERVED VALUES OF U IN THE MANN-WHITNEY TEST*

$n_2 = 3$

U \ n_1	1	2	3
0	.250	.100	.050
1	.500	.200	.100
2	.750	.400	.200
3		.600	.350
4			.500
5			.650

$n_2 = 4$

U \ n_1	1	2	3	4
0	.200	.067	.028	.014
1	.400	.133	.057	.029
2	.600	.267	.114	.057
3		.400	.200	.100
4		.600	.314	.171
5			.429	.243
6			.571	.343
7				.443
8				.557

$n_2 = 5$

U \ n_1	1	2	3	4	5
0	.167	.047	.018	.008	.004
1	.333	.095	.036	.016	.008
2	.500	.190	.071	.032	.016
3	.667	.286	.125	.056	.028
4		.429	.196	.095	.048
5		.571	.286	.143	.075
6			.393	.206	.111
7			.500	.278	.155
8			.607	.365	.210
9				.452	.274
10				.548	.345
11					.421
12					.500
13					.579

$n_2 = 6$

U \ n_1	1	2	3	4	5	6
0	.143	.036	.012	.005	.002	.001
1	.286	.071	.024	.010	.004	.002
2	.428	.143	.048	.019	.009	.004
3	.571	.214	.083	.033	.015	.008
4		.321	.131	.057	.026	.013
5		.429	.190	.086	.041	.021
6		.571	.274	.129	.063	.032
7			.357	.176	.089	.047
8			.452	.238	.123	.066
9			.548	.305	.165	.090
10				.381	.214	.120
11				.457	.268	.155
12				.545	.331	.197
13					.396	.242
14					.465	.294
15					.535	.350
16						.409
17						.469
18						.531

* Reproduced from Mann, H. B., and Whitney, D. R. 1947. On a test of whether one of two random variables is stochastically larger than the other. *Ann. Math. Statist.*, **18**, 52–54, with the kind permission of the authors and the publisher.

Table J. Table of Probabilities Associated with Values as Small as Observed Values of U in the Mann-Whitney Test* (*Continued*)

$$n_2 = 7$$

U \ n₁	1	2	3	4	5	6	7
0	.125	.028	.008	.003	.001	.001	.000
1	.250	.056	.017	.006	.003	.001	.001
2	.375	.111	.033	.012	.005	.002	.001
3	.500	.167	.058	.021	.009	.004	.002
4	.625	.250	.092	.036	.015	.007	.003
5		.333	.133	.055	.024	.011	.006
6		.444	.192	.082	.037	.017	.009
7		.556	.258	.115	.053	.026	.013
8			.333	.158	.074	.037	.019
9			.417	.206	.101	.051	.027
10			.500	.264	.134	.069	.036
11			.583	.324	.172	.090	.049
12				.394	.216	.117	.064
13				.464	.265	.147	.082
14				.538	.319	.183	.104
15					.378	.223	.130
16					.438	.267	.159
17					.500	.314	.191
18					.562	.365	.228
19						.418	.267
20						.473	.310
21						.527	.355
22							.402
23							.451
24							.500
25							.549

* Reproduced from Mann, H. B., and Whitney, D. R. 1947. On a test of whether one of two random variables is stochastically larger than the other. *Ann. Math. Statist.*, **18**, 52–54, with the kind permission of the authors and the publisher.

TABLE J. TABLE OF PROBABILITIES ASSOCIATED WITH VALUES AS SMALL AS OBSERVED VALUES OF U IN THE MANN-WHITNEY TEST* (*Continued*)

$n_2 = 8$

U	1	2	3	4	5	6	7	8	t	Normal
0	.111	.022	.006	.002	.001	.000	.000	.000	3.308	.001
1	.222	.044	.012	.004	.002	.001	.000	.000	3.203	.001
2	.333	.089	.024	.008	.003	.001	.001	.000	3.098	.001
3	.444	.133	.042	.014	.005	.002	.001	.001	2.993	.001
4	.556	.200	.067	.024	.009	.004	.002	.001	2.888	.002
5		.267	.097	.036	.015	.006	.003	.001	2.783	.003
6		.356	.139	.055	.023	.010	.005	.002	2.678	.004
7		.444	.188	.077	.033	.015	.007	.003	2.573	.005
8		.556	.248	.107	.047	.021	.010	.005	2.468	.007
9			.315	.141	.064	.030	.014	.007	2.363	.009
10			.387	.184	.085	.041	.020	.010	2.258	.012
11			.461	.230	.111	.054	.027	.014	2.153	.016
12			.539	.285	.142	.071	.036	.019	2.048	.020
13				.341	.177	.091	.047	.025	1.943	.026
14				.404	.217	.114	.060	.032	1.838	.033
15				.467	.262	.141	.076	.041	1.733	.041
16				.533	.311	.172	.095	.052	1.628	.052
17					.362	.207	.116	.065	1.523	.064
18					.416	.245	.140	.080	1.418	.078
19					.472	.286	.168	.097	1.313	.094
20					.528	.331	.198	.117	1.208	.113
21						.377	.232	.139	1.102	.135
22						.426	.268	.164	.998	.159
23						.475	.306	.191	.893	.185
24						.525	.347	.221	.788	.215
25							.389	.253	.683	.247
26							.433	.287	.578	.282
27							.478	.323	.473	.318
28							.522	.360	.368	.356
29								.399	.263	.396
30								.439	.158	.437
31								.480	.052	.481
32								.520		

* Reproduced from Mann, H. B., and Whitney, D. R. 1947. On a test of whether one of two random variables is stochastically larger than the other. *Ann. Math. Statist.*, **18**, 52–54, with the kind permission of the authors and the publisher.

TABLE K. TABLE OF CRITICAL VALUES OF U IN THE MANN-WHITNEY TEST*

Table K_I. Critical Values of U for a One-tailed Test at $\alpha = .001$ or for a Two-tailed Test at $\alpha = .002$

n_1 \ n_2	9	10	11	12	13	14	15	16	17	18	19	20
1												
2												
3									0	0	0	0
4		0	0	0	1	1	1	2	2	3	3	3
5	1	1	2	2	3	3	4	5	5	6	7	7
6	2	3	4	4	5	6	7	8	9	10	11	12
7	3	5	6	7	8	9	10	11	13	14	15	16
8	5	6	8	9	11	12	14	15	17	18	20	21
9	7	8	10	12	14	15	17	19	21	23	25	26
10	8	10	12	14	17	19	21	23	25	27	29	32
11	10	12	15	17	20	22	24	27	29	32	34	37
12	12	14	17	20	23	25	28	31	34	37	40	42
13	14	17	20	23	26	29	32	35	38	42	45	48
14	15	19	22	25	29	32	36	39	43	46	50	54
15	17	21	24	28	32	36	40	43	47	51	55	59
16	19	23	27	31	35	39	43	48	52	56	60	65
17	21	25	29	34	38	43	47	52	57	61	66	70
18	23	27	32	37	42	46	51	56	61	66	71	76
19	25	29	34	40	45	50	55	60	66	71	77	82
20	26	32	37	42	48	54	59	65	70	76	82	88

* Adapted and abridged from Tables 1, 3, 5, and 7 of Auble, D. 1953. Extended tables for the Mann-Whitney statistic. *Bulletin of the Institute of Educational Research at Indiana University*, **1**, No. 2, with the kind permission of the author and the publisher.

TABLE K. TABLE OF CRITICAL VALUES OF U IN THE MANN-WHITNEY TEST* (*Continued*)

Table K_{II}. Critical Values of U for a One-tailed Test at $\alpha = .01$ or for a Two-tailed Test at $\alpha = .02$

n_1 \ n_2	9	10	11	12	13	14	15	16	17	18	19	20
1												
2					0	0	0	0	0	0	1	1
3	1	1	1	2	2	2	3	3	4	4	4	5
4	3	3	4	5	5	6	7	7	8	9	9	10
5	5	6	7	8	9	10	11	12	13	14	15	16
6	7	8	9	11	12	13	15	16	18	19	20	22
7	9	11	12	14	16	17	19	21	23	24	26	28
8	11	13	15	17	20	22	24	26	28	30	32	34
9	14	16	18	21	23	26	28	31	33	36	38	40
10	16	19	22	24	27	30	33	36	38	41	44	47
11	18	22	25	28	31	34	37	41	44	47	50	53
12	21	24	28	31	35	38	42	46	49	53	56	60
13	23	27	31	35	39	43	47	51	55	59	63	67
14	26	30	34	38	43	47	51	56	60	65	69	73
15	28	33	37	42	47	51	56	61	66	70	75	80
16	31	36	41	46	51	56	61	66	71	76	82	87
17	33	38	44	49	55	60	66	71	77	82	88	93
18	36	41	47	53	59	65	70	76	82	88	94	100
19	38	44	50	56	63	69	75	82	88	94	101	107
20	40	47	53	60	67	73	80	87	93	100	107	114

* Adapted and abridged from Tables 1, 3, 5, and 7 of Auble, D. 1953. Extended tables for the Mann-Whitney statistic. *Bulletin of the Institute of Educational Research at Indiana University,* **1,** No. 2, with the kind permission of the author and the publisher.

TABLE K. TABLE OF CRITICAL VALUES OF U IN THE MANN-WHITNEY
TEST* (*Continued*)

Table K$_{\text{III}}$. Critical Values of U for a One-tailed Test at $\alpha = .025$ or for a Two-tailed
Test at $\alpha = .05$

n_1 \ n_2	9	10	11	12	13	14	15	16	17	18	19	20
1												
2	0	0	0	1	1	1	1	1	2	2	2	2
3	2	3	3	4	4	5	5	6	6	7	7	8
4	4	5	6	7	8	9	10	11	11	12	13	13
5	7	8	9	11	12	13	14	15	17	18	19	20
6	10	11	13	14	16	17	19	21	22	24	25	27
7	12	14	16	18	20	22	24	26	28	30	32	34
8	15	17	19	22	24	26	29	31	34	36	38	41
9	17	20	23	26	28	31	34	37	39	42	45	48
10	20	23	26	29	33	36	39	42	45	48	52	55
11	23	26	30	33	37	40	44	47	51	55	58	62
12	26	29	33	37	41	45	49	53	57	61	65	69
13	28	33	37	41	45	50	54	59	63	67	72	76
14	31	36	40	45	50	55	59	64	67	74	78	83
15	34	39	44	49	54	59	64	70	75	80	85	90
16	37	42	47	53	59	64	70	75	81	86	92	98
17	39	45	51	57	63	67	75	81	87	93	99	105
18	42	48	55	61	67	74	80	86	93	99	106	112
19	45	52	58	65	72	78	85	92	99	106	113	119
20	48	55	62	69	76	83	90	98	105	112	119	127

* Adapted and abridged from Tables 1, 3, 5, and 7 of Auble, D. 1953. Extended tables for the Mann-Whitney statistic. *Bulletin of the Institute of Educational Research at Indiana University*, **1**, No. 2, with the kind permission of the author and the publisher.

TABLE K. TABLE OF CRITICAL VALUES OF U IN THE MANN-WHITNEY
TEST* (*Continued*)

Table K$_{IV}$. Critical Values of U for a One-tailed Test at $\alpha = .05$ or for a Two-tailed
Test at $\alpha = .10$

n_2 \ n_1	9	10	11	12	13	14	15	16	17	18	19	20
1											0	0
2	1	1	1	2	2	2	3	3	3	4	4	4
3	3	4	5	5	6	7	7	8	9	9	10	11
4	6	7	8	9	10	11	12	14	15	16	17	18
5	9	11	12	13	15	16	18	19	20	22	23	25
6	12	14	16	17	19	21	23	25	26	28	30	32
7	15	17	19	21	24	26	28	30	33	35	37	39
8	18	20	23	26	28	31	33	36	39	41	44	47
9	21	24	27	30	33	36	39	42	45	48	51	54
10	24	27	31	34	37	41	44	48	51	55	58	62
11	27	31	34	38	42	46	50	54	57	61	65	69
12	30	34	38	42	47	51	55	60	64	68	72	77
13	33	37	42	47	51	56	61	65	70	75	80	84
14	36	41	46	51	56	61	66	71	77	82	87	92
15	39	44	50	55	61	66	72	77	83	88	94	100
16	42	48	54	60	65	71	77	83	89	95	101	107
17	45	51	57	64	70	77	83	89	96	102	109	115
18	48	55	61	68	75	82	88	95	102	109	116	123
19	51	58	65	72	80	87	94	101	109	116	123	130
20	54	62	69	77	84	92	100	107	115	123	130	138

* Adapted and abridged from Tables 1, 3, 5, and 7 of Auble, D. 1953. Extended tables for the Mann-Whitney statistic. *Bulletin of the Institute of Educational Research at Indiana University*, **1**, No. 2, with the kind permission of the author and the publisher.

TABLE L. TABLE OF CRITICAL VALUES OF K_D IN THE KOLMOGOROV-SMIRN0.
TWO-SAMPLE TEST
(Small samples)

N	One-tailed test*		Two-tailed test†	
	$\alpha = .05$	$\alpha = .01$	$\alpha = .05$	$\alpha = .01$
3	3	—	—	—
4	4	—	4	—
5	4	5	5	5
6	5	6	5	6
7	5	6	6	6
8	5	6	6	7
9	6	7	6	7
10	6	7	7	8
11	6	8	7	8
12	6	8	7	8
13	7	8	7	9
14	7	8	8	9
15	7	9	8	9
16	7	9	8	10
17	8	9	8	10
18	8	10	9	10
19	8	10	9	10
20	8	10	9	11
21	8	10	9	11
22	9	11	9	11
23	9	11	10	11
24	9	11	10	12
25	9	11	10	12
26	9	11	10	12
27	9	12	10	12
28	10	12	11	13
29	10	12	11	13
30	10	12	11	13
35	11	13	12	
40	11	14	13	

* Abridged from Goodman, L. A.   1954.   Kolmogorov-Smirnov tests for psychological research.   *Psychol. Bull.*, **51**, 167, with the kind permission of the author and the American Psychological Association.

† Derived from Table 1 of Massey, F. J., Jr.   1951.   The distribution of the maximum deviation between two sample cumulative step functions.   *Ann. Math. Statist.*, **22**, 126–127, with the kind permission of the author and the publisher.

TABLE M. TABLE OF CRITICAL VALUES OF D IN THE KOLMOGOROV-SMIRNOV
TWO-SAMPLE TEST
(Large samples: two-tailed test)*

Level of significance	Value of D so large as to call for rejection of H_0 at the indicated level of significance, where $D = \text{maximum } \lvert S_{n_1}(X) - S_{n_2}(X) \rvert$
.10	$1.22 \sqrt{\dfrac{n_1 + n_2}{n_1 n_2}}$
.05	$1.36 \sqrt{\dfrac{n_1 + n_2}{n_1 n_2}}$
.025	$1.48 \sqrt{\dfrac{n_1 + n_2}{n_1 n_2}}$
.01	$1.63 \sqrt{\dfrac{n_1 + n_2}{n_1 n_2}}$
.005	$1.73 \sqrt{\dfrac{n_1 + n_2}{n_1 n_2}}$
.001	$1.95 \sqrt{\dfrac{n_1 + n_2}{n_1 n_2}}$

* Adapted from Smirnov, N. 1948. Tables for estimating the goodness of fit of empirical distributions. *Ann. Math. Statist.*, **19**, 280–281, with the kind permission of the publisher.

TABLE N. TABLE OF PROBABILITIES ASSOCIATED WITH VALUES AS LARGE AS OBSERVED VALUES OF χ_r^2 IN THE FRIEDMAN TWO-WAY ANALYSIS OF VARIANCE BY RANKS*

Table N_I. $k = 3$

N = 2		N = 3		N = 4		N = 5	
χ_r^2	p	χ_r^2	p	χ_r^2	p	χ_r^2	p
0	1.000	.000	1.000	.0	1.000	.0	1.000
1	.833	.667	.944	.5	.931	.4	.954
3	.500	2.000	.528	1.5	.653	1.2	.691
4	.167	2.667	.361	2.0	.431	1.6	.522
		4.667	.194	3.5	.273	2.8	.367
		6.000	.028	4.5	.125	3.6	.182
				6.0	.069	4.8	.124
				6.5	.042	5.2	.093
				8.0	.0046	6.4	.039
						7.6	.024
						8.4	.0085
						10.0	.00077

N = 6		N = 7		N = 8		N = 9	
χ_r^2	p	χ_r^2	p	χ_r^2	p	χ_r^2	p
.00	1.000	.000	1.000	.00	1.000	.000	1.000
.33	.956	.286	.964	.25	.967	.222	.971
1.00	.740	.857	.768	.75	.794	.667	.814
1.33	.570	1.143	.620	1.00	.654	.889	.865
2.33	.430	2.000	.486	1.75	.531	1.556	.569
3.00	.252	2.571	.305	2.25	.355	2.000	.398
4.00	.184	3.429	.237	3.00	.285	2.667	.328
4.33	.142	3.714	.192	3.25	.236	2.889	.278
5.33	.072	4.571	.112	4.00	.149	3.556	.187
6.33	.052	5.429	.085	4.75	.120	4.222	.154
7.00	.029	6.000	.052	5.25	.079	4.667	.107
8.33	.012	7.143	.027	6.25	.047	5.556	.069
9.00	.0081	7.714	.021	6.75	.038	6.000	.057
9.33	.0055	8.000	.016	7.00	.030	6.222	.048
10.33	.0017	8.857	.0084	7.75	.018	6.889	.031
12.00	.00013	10.286	.0036	9.00	.0099	8.000	.019
		10.571	.0027	9.25	.0080	8.222	.016
		11.143	.0012	9.75	.0048	8.667	.010
		12.286	.00032	10.75	.0024	9.556	.0060
		14.000	.000021	12.00	.0011	10.667	.0035
				12.25	.00086	10.889	.0029
				13.00	.00026	11.556	.0013
				14.25	.000061	12.667	.00066
				16.00	.0000036	13.556	.00035
						14.000	.00020
						14.222	.000097
						14.889	.000054
						16.222	.000011
						18.000	.0000006

* Adapted from Friedman, M. 1937. The use of ranks to avoid the assumption of normality implicit in the analysis of variance. *J. Amer. Statist. Ass.*, **32**, 688–689, with the kind permission of the author and the publisher.

Table N. Table of Probabilities Associated with Values as Large as Observed Values of χ_r^2 in the Friedman Two-way Analysis of Variance by Ranks* (*Continued*)

Table N$_{\text{II}}$. $k = 4$

$N = 2$		$N = 3$		$N = 4$			
χ_r^2	p	χ_r^2	p	χ_r^2	p	χ_r^2	p
.0	1.000	.2	1.000	.0	1.000	5.7	.141
.6	.958	.6	.958	.3	.992	6.0	.105
1.2	.834	1.0	.910	.6	.928	6.3	.094
1.8	.792	1.8	.727	.9	.900	6.6	.077
2.4	.625	2.2	.608	1.2	.800	6.9	.068
3.0	.542	2.6	.524	1.5	.754	7.2	.054
3.6	.458	3.4	.446	1.8	.677	7.5	.052
4.2	.375	3.8	.342	2.1	.649	7.8	.036
4.8	.208	4.2	.300	2.4	.524	8.1	.033
5.4	.167	5.0	.207	2.7	.508	8.4	.019
6.0	.042	5.4	.175	3.0	.432	8.7	.014
		5.8	.148	3.3	.389	9.3	.012
		6.6	.075	3.6	.355	9.6	.0069
		7.0	.054	3.9	.324	9.9	.0062
		7.4	.033	4.5	.242	10.2	.0027
		8.2	.017	4.8	.200	10.8	.0016
		9.0	.0017	5.1	.190	11.1	.00094
				5.4	.158	12.0	.000072

* Adapted from Friedman, M. 1937. The use of ranks to avoid the assumption of normality implicit in the analysis of variance. *J. Amer. Statist. Ass.*, **32**, 688–689, with the kind permission of the author and the publisher.

TABLE O. TABLE OF PROBABILITIES ASSOCIATED WITH VALUES AS LARGE AS OBSERVED VALUES OF H IN THE KRUSKAL-WALLIS ONE-WAY ANALYSIS OF VARIANCE BY RANKS*

Sample sizes			H	p	Sample sizes			H	p
n_1	n_2	n_3			n_1	n_2	n_3		
2	1	1	2.7000	.500	4	3	2	6.4444	.008
								6.3000	.011
2	2	1	3.6000	.200				5.4444	.046
								5.4000	.051
2	2	2	4.5714	.067				4.5111	.098
			3.7143	.200				4.4444	.102
3	1	1	3.2000	.300	4	3	3	6.7455	.010
3	2	1	4.2857	.100				6.7091	.013
			3.8571	.133				5.7909	.046
3	2	2	5.3572	.029				5.7273	.050
			4.7143	.048				4.7091	.092
			4.5000	.067				4.7000	.101
			4.4643	.105	4	4	1	6.6667	.010
3	3	1	5.1429	.043				6.1667	.022
			4.5714	.100				4.9667	.048
			4.0000	.129				4.8667	.054
3	3	2	6.2500	.011				4.1667	.082
			5.3611	.032				4.0667	.102
			5.1389	.061	4	4	2	7.0364	.006
			4.5556	.100				6.8727	.011
			4.2500	.121				5.4545	.046
3	3	3	7.2000	.004				5.2364	.052
			6.4889	.011				4.5545	.098
			5.6889	.029				4.4455	.103
			5.6000	.050	4	4	3	7.1439	.010
			5.0667	.086				7.1364	.011
			4.6222	.100				5.5985	.049
4	1	1	3.5714	.200				5.5758	.051
4	2	1	4.8214	.057				4.5455	.099
			4.5000	.076				4.4773	.102
			4.0179	.114	4	4	4	7.6538	.008
4	2	2	6.0000	.014				7.5385	.011
			5.3333	.033				5.6923	.049
			5.1250	.052				5.6538	.054
			4.4583	.100				4.6539	.097
			4.1667	.105				4.5001	.104
4	3	1	5.8333	.021	5	1	1	3.8571	.143
			5.2083	.050	5	2	1	5.2500	.036
			5.0000	.057				5.0000	.048
			4.0556	.093				4.4500	.071
			3.8889	.129				4.2000	.095
								4.0500	.119

TABLE O. TABLE OF PROBABILITIES ASSOCIATED WITH VALUES AS LARGE AS OBSERVED VALUES OF H IN THE KRUSKAL-WALLIS ONE-WAY ANALYSIS OF VARIANCE BY RANKS* (*Continued*)

Sample sizes			H	p	Sample sizes			H	p
n_1	n_2	n_3			n_1	n_2	n_3		
5	2	2	6.5333	.008				5.6308	.050
			6.1333	.013				4.5487	.099
			5.1600	.034				4.5231	.103
			5.0400	.056	5	4	4	7.7604	.009
			4.3733	.090				7.7440	.011
			4.2933	.122				5.6571	.049
								5.6176	.050
5	3	1	6.4000	.012				4.6187	.100
			4.9600	.048				4.5527	.102
			4.8711	.052					
			4.0178	.095	5	5	1	7.3091	.009
			3.8400	.123				6.8364	.011
								5.1273	.046
5	3	2	6.9091	.009				4.9091	.053
			6.8218	.010				4.1091	.086
			5.2509	.049				4.0364	.105
			5.1055	.052					
			4.6509	.091	5	5	2	7.3385	.010
			4.4945	.101				7.2692	.010
								5.3385	.047
5	3	3	7.0788	.009				5.2462	.051
			6.9818	.011				4.6231	.097
			5.6485	.049				4.5077	.100
			5.5152	.051					
			4.5333	.097	5	5	3	7.5780	.010
			4.4121	.109				7.5429	.010
								5.7055	.046
5	4	1	6.9545	.008				5.6264	.051
			6.8400	.011				4.5451	.100
			4.9855	.044				4.5363	.102
			4.8600	.056					
			3.9873	.098	5	5	4	7.8229	.010
			3.9600	.102				7.7914	.010
								5.6657	.049
5	4	2	7.2045	.009				5.6429	.050
			7.1182	.010				4.5229	.099
			5.2727	.049				4.5200	.101
			5.2682	.050					
			4.5409	.098	5	5	5	8.0000	.009
			4.5182	.101				7.9800	.010
								5.7800	.049
5	4	3	7.4449	.010				5.6600	.051
			7.3949	.011				4.5600	.100
			5.6564	.049				4.5000	.102

* Adapted and abridged from Kruskal, W. H., and Wallis, W. A. 1952. Use of ranks in one-criterion variance analysis. *J. Amer. Statist. Ass.*, **47**, 614–617, with the kind permission of the authors and the publisher. (The corrections to this table given by the authors in Errata, *J. Amer. Statist. Ass.*, **48**, 910, have been incorporated.)

TABLE P. TABLE OF CRITICAL VALUES OF r_S, THE SPEARMAN RANK
CORRELATION COEFFICIENT*

N	Significance level (one-tailed test)	
	.05	.01
4	1.000	
5	.900	1.000
6	.829	.943
7	.714	.893
8	.643	.833
9	.600	.783
10	.564	.746
12	.506	.712
14	.456	.645
16	.425	.601
18	.399	.564
20	.377	.534
22	359	.508
24	.343	.485
26	.329	.465
28	.317	.448
30	.306	.432

* Adapted from Olds, E. G. 1938. Distributions of sums of squares of rank differences for small numbers of individuals. *Ann. Math. Statist.*, **9**, 133–148, and from Olds, E. G. 1949. The 5% significance levels for sums of squares of rank differences and a correction. *Ann. Math. Statist.*, **20**, 117–118, with the kind permission of the author and the publisher.

TABLE Q. TABLE OF PROBABILITIES ASSOCIATED WITH VALUES AS LARGE AS OBSERVED VALUES OF S IN THE KENDALL RANK CORRELATION COEFFICIENT

S	Values of N				S	Values of N		
	4	5	8	9		6	7	10
0	.625	.592	.548	.540	1	.500	.500	.500
2	.375	.408	.452	.460	3	.360	.386	.431
4	.167	.242	.360	.381	5	.235	.281	.364
6	.042	.117	.274	.306	7	.136	.191	.300
8		.042	.199	.238	9	.068	.119	.242
10		.0083	.138	.179	11	.028	.068	.190
12			.089	.130	13	.0083	.035	.146
14			.054	.090	15	.0014	.015	.108
16			.031	.060	17		.0054	.078
18			.016	.038	19		.0014	.054
20			.0071	.022	21		.00020	.036
22			.0028	.012	23			.023
24			.00087	.0063	25			.014
26			.00019	.0029	27			.0083
28			.000025	.0012	29			.0046
30				.00043	31			.0023
32				.00012	33			.0011
34				.000025	35			.00047
36				.0000028	37			.00018
					39			.000058
					41			.000015
					43			.0000028
					45			.00000028

* Adapted by permission from Kendall, M. G., *Rank correlation methods*, Charles Griffin & Company, Ltd., London, 1948, Appendix Table 1 p. 141.

TABLE R. TABLE OF CRITICAL VALUES OF s IN THE KENDALL COEFFICIENT
OF CONCORDANCE*

k	N					Additional values for $N = 3$	
	3†	4	5	6	7	k	s
Values at the .05 level of significance							
3			64.4	103.9	157.3	9	54.0
4		49.5	88.4	143.3	217.0	12	71.9
5		62.6	112.3	182.4	276.2	14	83.8
6		75.7	136.1	221.4	335.2	16	95.8
8	48.1	101.7	183.7	299.0	453.1	18	107.7
10	60.0	127.8	231.2	376.7	571.0		
15	89.8	192.9	349.8	570.5	864.9		
20	119.7	258.0	468.5	764.4	1,158.7		
Values at the .01 level of significance							
3			75.6	122.8	185.6	9	75.9
4		61.4	109.3	176.2	265.0	12	103.5
5		80.5	142.8	229.4	343.8	14	121.9
6		99.5	176.1	282.4	422.6	16	140.2
8	66.8	137.4	242.7	388.3	579.9	18	158.6
10	85.1	175.3	309.1	494.0	737.0		
15	131.0	269.8	475.2	758.2	1,129.5		
20	177.0	364.2	641.2	1,022.2	1,521.9		

* Adapted from Friedman, M. 1940. A comparison of alternative tests of significance for the problem of m rankings. *Ann. Math. Statist.*, **11**, 86–92, with the kind permission of the author and the publisher.

† Notice that additional critical values of s for $N = 3$ are given in the right-hand column of this table.

Table S. Table of Factorials

N	$N!$
0	1
1	1
2	2
3	6
4	24
5	120
6	720
7	5040
8	40320
9	362880
10	3628800
11	39916800
12	479001600
13	6227020800
14	87178291200
15	1307674368000
16	20922789888000
17	355687428096000
18	6402373705728000
19	121645100408832000
20	2432902008176640000

TABLE T. TABLE OF BINOMIAL COEFFICIENTS

N	$\binom{N}{0}$	$\binom{N}{1}$	$\binom{N}{2}$	$\binom{N}{3}$	$\binom{N}{4}$	$\binom{N}{5}$	$\binom{N}{6}$	$\binom{N}{7}$	$\binom{N}{8}$	$\binom{N}{9}$	$\binom{N}{10}$
0	1										
1	1	1									
2	1	2	1								
3	1	3	3	1							
4	1	4	6	4	1						
5	1	5	10	10	5	1					
6	1	6	15	20	15	6	1				
7	1	7	21	35	35	21	7	1			
8	1	8	28	56	70	56	28	8	1		
9	1	9	36	84	126	126	84	36	9	1	
10	1	10	45	120	210	252	210	120	45	10	1
11	1	11	55	165	330	462	462	330	165	55	11
12	1	12	66	220	495	792	924	792	495	220	66
13	1	13	78	286	715	1287	1716	1716	1287	715	286
14	1	14	91	364	1001	2002	3003	3432	3003	2002	1001
15	1	15	105	455	1365	3003	5005	6435	6435	5005	3003
16	1	16	120	560	1820	4368	8008	11440	12870	11440	8008
17	1	17	136	680	2380	6188	12376	19448	24310	24310	19448
18	1	18	153	816	3060	8568	18564	31824	43758	48620	43758
19	1	19	171	969	3876	11628	27132	50388	75582	92378	92378
20	1	20	190	1140	4845	15504	38760	77520	125970	167960	184756

TABLE U. TABLE OF SQUARES AND SQUARE ROOTS*

Number	Square	Square root	Number	Square	Square root
1	1	1.0000	41	16 81	6.4031
2	4	1.4142	42	17 64	6.4807
3	9	1.7321	43	18 49	6.5574
4	16	2.0000	44	19 36	6.6332
5	25	2.2361	45	20 25	6.7082
6	36	2.4495	46	21 16	6.7823
7	49	2.6458	47	22 09	6.8557
8	64	2.8284	48	23 04	6.9282
9	81	3.0000	49	24 01	7.0000
10	1 00	3.1623	50	25 00	7.0711
11	1 21	3.3166	51	26 01	7.1414
12	1 44	3.4641	52	27 04	7.2111
13	1 69	3.6056	53	28 09	7.2801
14	1 96	3.7417	54	29 16	7.3485
15	2 25	3.8730	55	30 25	7.4162
16	2 56	4.0000	56	31 36	7.4833
17	2 89	4.1231	57	32 49	7.5498
18	3 24	4.2426	58	33 64	7.6158
19	3 61	4.3589	59	34 81	7.6811
20	4 00	4.4721	60	36 00	7.7460
21	4 41	4.5826	61	37 21	7.8102
22	4 84	4.6904	62	38 44	7.8740
23	5 29	4.7958	63	39 69	7.9373
24	5 76	4.8990	64	40 96	8.0000
25	6 25	5.0000	65	42 25	8.0623
26	6 76	5.0990	66	43 56	8.1240
27	7 29	5.1962	67	44 89	8.1854
28	7 84	5.2915	68	46 24	8.2462
29	8 41	5.3852	69	47 61	8.3066
30	9 00	5.4772	70	49 00	8.3666
31	9 61	5.5678	71	50 41	8.4261
32	10 24	5.6569	72	51 84	8.4853
33	10 89	5.7446	73	53 29	8.5440
34	11 56	5.8310	74	54 76	8.6023
35	12 25	5.9161	75	56 25	8.6603
36	12 96	6.0000	76	57 76	8.7178
37	13 69	6.0828	77	59 29	8.7750
38	14 44	6.1644	78	60 84	8.8318
39	15 21	6.2450	79	62 41	8.8882
40	16 00	6.3246	80	64 00	8.9443

* By permission from *Statistics for students of psychology and education,* by H. Sorenson.   Copyright 1936, McGraw-Hill Book Company, Inc.

TABLE U. TABLE OF SQUARES AND SQUARE ROOTS (*Continued*)

Number	Square	Square root	Number	Square	Square root
81	65 61	9.0000	121	1 46 41	11.0000
82	67 24	9.0554	122	1 48 84	11.0454
83	68 89	9.1104	123	1 51 29	11.0905
84	70 56	9.1652	124	1 53 76	11.1355
85	72 25	9.2195	125	1 56 25	11.1803
86	73 96	9.2736	126	1 58 76	11.2250
87	75 69	9.3274	127	1 61 29	11.2694
88	77 44	9.3808	128	1 63 84	11.3137
89	79 21	9.4340	129	1 66 41	11.3578
90	81 00	9.4868	130	1 69 00	11.4018
91	82 81	9.5394	131	1 71 61	11.4455
92	84 64	9.5917	132	1 74 24	11.4891
93	86 49	9.6437	133	1 76 89	11.5326
94	88 36	9.6954	134	1 79 56	11.5758
95	90 25	9.7468	135	1 82 25	11.6190
96	92 16	9.7980	136	1 84 96	11.6619
97	94 09	9.8489	137	1 87 69	11.7047
98	96 04	9.8995	138	1 90 44	11.7473
99	98 01	9.9499	139	1 93 21	11.7898
100	1 00 00	10.0000	140	1 96 00	11.8322
101	1 02 01	10.0499	141	1 98 81	11.8743
102	1 04 04	10.0995	142	2 01 64	11.9164
103	1 06 09	10.1489	143	2 04 49	11.9583
104	1 08 16	10.1980	144	2 07 36	12.0000
105	1 10 25	10.2470	145	2 10 25	12.0416
106	1 12 36	10.2956	146	2 13 16	12.0830
107	1 14 49	10.3441	147	2 16 09	12.1244
108	1 16 64	10.3923	148	2 19 04	12.1655
109	1 18 81	10.4403	149	2 22 01	12.2066
110	1 21 00	10.4881	150	2 25 00	12.2474
111	1 23 21	10.5357	151	2 28 01	12.2882
112	1 25 44	10.5830	152	2 31 04	12.3288
113	1 27 69	10.6301	153	2 34 09	12.3693
114	1 29 96	10.6771	154	2 37 16	12.4097
115	1 32 25	10.7238	155	2 40 25	12.4499
116	1 34 56	10.7703	156	2 43 36	12.4900
117	1 36 89	10.8167	157	2 46 49	12.5300
118	1 39 24	10.8628	158	2 49 64	12.5698
119	1 41 61	10.9087	159	2 52 81	12.6095
120	1 44 00	10.9545	160	2 56 00	12.6491

Table U. Table of Squares and Square Roots* *(Continued)*

Number	Square	Square root	Number	Square	Square root
161	2 59 21	12.6886	201	4 04 01	14.1774
162	2 62 44	12.7279	202	4 08 04	14.2127
163	2 65 69	12.7671	203	4 12 09	14.2478
164	2 68 96	12.8062	204	4 16 16	14.2829
165	2 72 25	12.8452	205	4 20 25	14.3178
166	2 75 56	12.8841	206	4 24 36	14.3527
167	2 78 89	12.9228	207	4 28 49	14.3875
168	2 82 24	12.9615	208	4 32 64	14.4222
169	2 85 61	13.0000	209	4 36 81	14.4568
170	2 89 00	13.0384	210	4 41 00	14.4914
171	2 92 41	13.0767	211	4 45 21	14.5258
172	2 95 84	13.1149	212	4 49 44	14.5602
173	2 99 29	13.1529	213	4 53 69	14.5945
174	3 02 76	13.1909	214	4 57 96	14.6287
175	3 06 25	13.2288	215	4 62 25	14.6629
176	3 09 76	13.2665	216	4 66 56	14.6969
177	3 13 29	13.3041	217	4 70 89	14.7309
178	3 16 84	13.3417	218	4 75 24	14.7648
179	3 20 41	13.3791	219	4 79 61	14.7986
180	3 24 00	13.4164	220	4 84 00	14.8324
181	3 27 61	13.4536	221	4 88 41	14.8661
182	3 31 24	13.4907	222	4 92 84	14.8997
183	3 34 89	13.5277	223	4 97 29	14.9332
184	3 38 56	13.5647	224	5 01 76	14.9666
185	3 42 25	13.6015	225	5 06 25	15.0000
186	3 45 96	13.6382	226	5 10 76	15.0333
187	3 49 69	13.6748	227	5 15 29	15.0665
188	3 53 44	13.7113	228	5 19 84	15.0997
189	3 57 21	13.7477	229	5 24 41	15.1327
190	3 61 00	13.7840	230	5 29 00	15.1658
191	3 64 81	13.8203	231	5 33 61	15.1987
192	3 68 64	13.8564	232	5 38 24	15.2315
193	3 72 49	13.8924	233	5 42 89	15.2643
194	3 76 36	13.9284	234	5 47 56	15.2971
195	3 80 25	13.9642	235	5 52 25	15.3297
196	3 84 16	14.0000	236	5 56 96	15.3623
197	3 88 09	14.0357	237	5 61 69	15.3948
198	3 92 04	14.0712	238	5 66 44	15.4272
199	3 96 01	14.1067	239	5 71 21	15.4596
200	4 00 00	14.1421	240	5 76 00	15.4919

* By permission from *Statistics for students of psychology and education*, by H. Sorenson. Copyright 1936, McGraw-Hill Book Company, Inc.

TABLE U. TABLE OF SQUARES AND SQUARE ROOTS* (Continued)

Number	Square	Square root	Number	Square	Square root
241	5 80 81	15.5242	281	7 89 61	16.7631
242	5 85 64	15.5563	282	7 95 24	16.7929
243	5 90 49	15.5885	283	8 00 89	16.8226
244	5 95 36	15.6205	284	8 06 56	16.8523
245	6 00 25	15.6525	285	8 12 25	16.8819
246	6 05 16	15.6844	286	8 17 96	16.9115
247	6 10 09	15.7162	287	8 23 69	16.9411
248	6 15 04	15.7480	288	8 29 44	16.9706
249	6 20 01	15.7797	289	8 35 21	17.0000
250	6 25 00	15.8114	290	8 41 00	17.0294
251	6 30 01	15.8430	291	8 46 81	17.0587
252	6 35 04	15.8745	292	8 52 64	17.0880
253	6 40 09	15.9060	293	8 58 49	17.1172
254	6 45 16	15.9374	294	8 64 36	17.1464
255	6 50 25	15.9687	295	8 70 25	17.1756
256	6 55 36	16.0000	296	8 76 16	17.2047
257	6 60 49	16.0312	297	8 82 09	17.2337
258	6 65 64	16.0624	298	8 88 04	17.2627
259	6 70 81	16.0935	299	8 94 01	17.2916
260	6 76 00	16.1245	300	9 00 00	17.3205
261	6 81 21	16.1555	301	9 06 01	17.3494
262	6 86 44	16.1864	302	9 12 04	17.3781
263	6 91 69	16.2173	303	9 18 09	17.4069
264	6 96 96	16.2481	304	9 24 16	17.4356
265	7 02 25	16.2788	305	9 30 25	17.4642
266	7 07 56	16.3095	306	9 36 36	17.4929
267	7 12 89	16.3401	307	9 42 49	17.5214
268	7 18 24	16.3707	308	9 48 64	17.5499
269	7 23 61	16.4012	309	9 54 81	17.5784
270	7 29 00	16.4317	310	9 61 00	17.6068
271	7 34 41	16.4621	311	9 67 21	17.6352
272	7 39 84	16.4924	312	9 73 44	17.6635
273	7 45 29	16.5227	313	9 79 69	17.6918
274	7 50 76	16.5529	314	9 85 96	17.7200
275	7 56 25	16.5831	315	9 92 25	17.7482
276	7 61 76	16.6132	316	9 98 56	17.7764
277	7 67 29	16.6433	317	10 04 89	17.8045
278	7 72 84	16.6733	318	10 11 24	17.8326
279	7 78 41	16.7033	319	10 17 61	17.8606
280	7 84 00	16.7332	320	10 24 00	17.8885

* By permission from Statistics for students of psychology and education, by H. Sorenson. Copyright 1936, McGraw-Hill Book Company, Inc.

TABLE U. TABLE OF SQUARES AND SQUARE ROOTS* (*Continued*)

Number	Square	Square root	Number	Square	Square root
321	10 30 41	17.9165	361	13 03 21	19.0000
322	10 36 84	17.9444	362	13 10 44	19.0263
323	10 43 29	17.9722	363	13 17 69	19.0526
324	10 49 76	18.0000	364	13 24 96	19.0788
325	10 56 25	18.0278	365	13 32 25	19.1050
326	10 62 76	18.0555	366	13 39 56	19.1311
327	10 69 29	18.0831	367	13 46 89	19.1572
328	10 75 84	18.1108	368	13 54 24	19.1833
329	10 82 41	18.1384	369	13 61 61	19.2094
330	10 89 00	18.1659	370	13 69 00	19.2354
331	10 95 61	18.1934	371	13 76 41	19.2614
332	11 02 24	18.2209	372	13 83 84	19.2873
333	11 08 89	18.2483	373	13 91 29	19.3132
334	11 15 56	18.2757	374	13 98 76	19.3391
335	11 22 25	18.3030	375	14 06 25	19.3649
336	11 28 96	18.3303	376	14 13 76	19.3907
337	11 35 69	18.3576	377	14 21 29	19.4165
338	11 42 44	18.3848	378	14 28 84	19.4422
339	11 49 21	18.4120	379	14 36 41	19.4679
340	11 56 00	18.4391	380	14 44 00	19.4936
341	11 62 81	18.4662	381	14 51 61	19.5192
342	11 69 64	18.4932	382	14 59 24	19.5448
343	11 76 49	18.5203	383	14 66 89	19.5704
344	11 83 36	18.5472	384	14 74 56	19.5959
345	11 90 25	18.5742	385	14 82 25	19.6214
346	11 97 16	18.6011	386	14 89 96	19.6469
347	12 04 09	18.6279	387	14 97 69	19.6723
348	12 11 04	18.6548	388	15 05 44	19.6977
349	12 18 01	18.6815	389	15 13 21	19.7231
350	12 25 00	18.7083	390	15 21 00	19.7484
351	12 32 01	18.7350	391	15 28 81	19.7737
352	12 39 04	18.7617	392	15 36 64	19.7990
353	12 46 09	18.7883	393	15 44 49	19.8242
354	12 53 16	18.8149	394	15 52 36	19.8494
355	12 60 25	18.8414	395	15 60 25	19.8746
356	12 67 36	18.8680	396	15 68 16	19.8997
357	12 74 49	18.8944	397	15 76 09	19.9249
358	12 81 64	18.9209	398	15 84 04	19.9499
359	12 88 81	18.9473	399	15 92 01	19.9750
360	12 96 00	18.9737	400	16 00 00	20.0000

* By permission from *Statistics for students of psychology and education,* by H. Sorenson. Copyright 1936, McGraw-Hill Book Company, Inc.

TABLE U. TABLE OF SQUARES AND SQUARE ROOTS* (*Continued*)

Number	Square	Square root	Number	Square	Square root
401	16 08 01	20.0250	441	19 44 81	21.0000
402	16 16 04	20.0499	442	19 53 64	21.0238
403	16 24 09	20.0749	443	19 62 49	'21.0476
404	16 32 16	20.0998	444	19 71 36	21.0713
405	16 40 25	20.1246	445	19 80 25	21.0950
406	16 48 36	20.1494	446	19 89 16	21.1187
407	16 56 49	20.1742	447	19 98 09	21.1424
408	16 64 64	20.1990	448	20 07 04	21.1660
409	16 72 81	20.2237	449	20 16 01	21.1896
410	16 81 00	20.2485	450	20 25 00	21.2132
411	16 89 21	20.2731	451	20 34 01	21.2368
412	16 97 44	20.2978	452	20 43 04	21.2603
413	17 05 69	20.3224	453	20 52 09	21.2838
414	17 13 96	20.3470	454	20 61 16	21.3073
415	17 22 25	20.3715	455	20 70 25	21.3307
416	17 30 56	20.3961	456	20 79 36	21.3542
417	17 38 89	20.4206	457	20 88 49	21.3776
418	17 47 24	20.4450	458	20 97 64	21.4009
419	17 55 61	20.4695	459	21 06 81	21.4243
420	17 64 00	20.4939	460	21 16 00	21.4476
421	17 72 41	20.5183	461	21 25 21	21.4709
422	17 80 84	20.5426	462	21 34 44	21.4942
423	17 89 29	20.5670	463	21 43 69	21.5174
424	17 97 76	20.5913	464	21 52 96	21.5407
425	18 06 25	20.6155	465	21 62 25	21.5639
426	18 14 76	20.6398	466	21 71 56	21.5870
427	18 23 29	20.6640	467	21 80 89	21.6102
428	18 31 84	20.6882	468	21 90 24	21.6333
429	18 40 41	20.7123	469	21 99 61	21.6564
430	18 49 00	20.7364	470	22 09 00	21.6795
431	18 57 61	20.7605	471	22 18 41	21.7025
432	18 66 24	20.7846	472	22 27 84	21.7256
433	18 74 89	20.8087	473	22 37 29	21.7486
434	18 83 56	20.8327	474	22 46 76	21.7715
435	18 92 25	20.8567	475	22 56 25	21.7945
436	19 00 96	20.8806	476	22 65 76	21.8174
437	19 09 69	20.9045	477	22 75 29	21.8403
438	19 18 44	20.9284	478	22 84 84	21.8632
439	19 27 21	20.9523	479	22 94 41	21.8861
440	19 36 00	20.9762	480	23 04 00	21.9089

* By permission from *Statistics for students of psychology and education*, by H. Sorenson. Copyright 1936, McGraw-Hill Book Company, Inc.

TABLE U. TABLE OF SQUARES AND SQUARE ROOTS* (*Continued*)

Number	Square	Square root	Number	Square	Square root
481	23 13 61	21.9317	521	27 14 41	22.8254
482	23 23 24	21.9545	522	27 24 84	22.8473
483	23 32 89	21.9773	523	27 35 29	22.8692
484	23 42 56	22.0000	524	27 45 76	22.8910
485	23 52 25	22.0227	525	27 56 25	22.9129
486	23 61 96	22.0454	526	27 66 76	22.9347
487	23 71 69	22.0681	527	27 77 29	22.9565
488	23 81 44	22.0907	528	27 87 84	22.9783
489	23 91 21	22.1133	529	27 98 41	23.0000
490	24 01 00	22.1359	530	28 09 00	23.0217
491	24 10 81	22.1585	531	28 19 61	23.0434
492	24 20 64	22.1811	532	28 30 24	23.0651
493	24 30 49	22.2036	533	28 40 89	23.0868
494	24 40 36	22.2261	534	28 51 56	23.1084
495	24 50 25	22.2486	535	28 62 25	23.1301
496	24 60 16	22.2711	536	28 72 96	23.1517
497	24 70 09	22.2935	537	28 83 69	23.1733
498	24 80 04	22.3159	538	28 94 44	23.1948
499	24 90 01	22.3383	539	29 05 21	23.2164
500	25 00 00	22.3607	540	29 16 00	23.2379
501	25 10 01	22.3830	541	29 26 81	23.2594
502	25 20 04	22.4054	542	29 37 64	23.2809
503	25 30 09	22.4277	543	29 48 49	23.3024
504	25 40 16	22.4499	544	29 59 36	23.3238
505	25 50 25	22.4722	545	29 70 25	23.3452
506	25 60 36	22.4944	546	29 81 16	23.3666
507	25 70 49	22.5167	547	29 92 09	23.3880
508	25 80 64	22.5389	548	30 03 04	23.4094
509	25 90 81	22.5610	549	30 14 01	23.4307
510	26 01 00	22.5832	550	30 25 00	23.4521
511	26 11 21	22.6053	551	30 36 01	23.4734
512	26 21 44	22.6274	552	30 47 04	23.4947
513	26 31 69	22.6495	553	30 58 09	23.5160
514	26 41 96	22.6716	554	30 69 16	23.5372
515	26 52 25	22.6936	555	30 80 25	23.5584
516	26 62 56	22.7156	556	30 91 36	23.5797
517	26 72 89	22.7376	557	31 02 49	23.6008
518	26 83 24	22.7596	558	31 13 64	23.6220
519	26 93 61	22.7816	559	31 24 81	23.6432
520	27 04 00	22.8035	560	31 36 00	23.6643

* By permission from *Statistics for students of psychology and education*, by H. Sorenson. Copyright 1936, McGraw-Hill Book Company, Inc.

TABLE U. TABLE OF SQUARES AND SQUARE ROOTS* (*Continued*)

Number	Square	Square root	Number	Square	Square root
561	31 47 21	23.6854	601	36 12 01	24.5153
562	31 58 44	23.7065	602	36 24 04	24.5357
563	31 69 69	23.7276	603	36 36 09	24.5561
564	31 80 96	23.7487	604	36 48 16	24.5764
565	31 92 25	23.7697	605	36 60 25	24.5967
566	32 03 56	23.7908	606	36 72 36	24.6171
567	32 14 89	23.8118	607	36 84 49	24.6374
568	32 26 24	23.8328	608	36 96 64	24.6577
569	32 37 61	23.8537	609	37 08 81	24.6779
570	32 49 00	23.8747	610	37 21 00	24.6982
571	32 60 41	23.8956	611	37 33 21	24.7184
572	32 71 84	23.9165	612	37 45 44	24.7385
573	32 83 29	23.9374	613	37 57 69	24.7588
574	32 94 76	23.9583	614	37 69 96	24.7790
575	33 06 25	23.9792	615	37 82 25	24.7992
576	33 17 76	24.0000	616	37 94 56	24.8193
577	33 29 29	24.0208	617	38 06 89	24.8395
578	33 40 84	24.0416	618	38 19 24	24.8596
579	33 52 41	24.0624	619	38 31 61	24.8797
580	33 64 00	24.0832	620	38 44 00	24.8998
581	33 75 61	24.1039	621	38 56 41	24.9199
582	33 87 24	24.1247	622	38 68 84	24.9399
583	33 98 89	24.1454	623	38 81 29	24.9600
584	34 10 56	24.1661	624	38 93 76	24.9800
585	34 22 25	24.1868	625	39 06 25	25.0000
586	34 33 96	24.2074	626	39 18 76	25.0200
587	34 45 69	24.2281	627	39 31 29	25.0400
588	34 57 44	24.2487	628	39 43 84	25.0599
589	34 69 21	24.2693	629	39 56 41	25.0799
590	34 81 00	24.2899	630	39 69 00	25.0998
591	34 92 81	24.3105	631	39 81 61	25.1197
592	35 04 64	24.3311	632	39 94 24	25.1396
593	35 16 49	24.3516	633	40 06 89	25.1595
594	35 28 36	24.3721	634	40 19 56	25.1794
595	35 40 25	24.3926	635	40 32 25	25.1992
596	35 52 16	24.4131	636	40 44 96	25.2190
597	35 64 09	24.4336	637	40 57 69	25.2389
598	35 76 04	24.4540	638	40 70 44	25.2587
599	35 88 01	24.4745	639	40 83 21	25.2784
600	36 00 00	24.4949	640	40 96 00	25.2982

TABLE U. TABLE OF SQUARES AND SQUARE ROOTS* (*Continued*)

Number	Square	Square root	Number	Square	Square root
641	41 08 81	25.3180	681	46 37 61	26.0960
642	41 21 64	25.3377	682	46 51 24	26.1151
643	41 34 49	25.3574	683	46 64 89	26.1343
644	41 47 36	25.3772	684	46 78 56	26.1534
645	41 60 25	25.3969	685	46 92 25	26.1725
646	41 73 16	25.4165	686	47 05 96	26.1916
647	41 86 09	25.4362	687	47 19 69	26.2107
648	41 99 04	25.4558	688	47 33 44	26.2298
649	42 12 01	25.4755	689	47 47 21	26.2488
650	42 25 00	25.4951	690	47 61 00	26.2679
651	42 38 01	25.5147	691	47 74 81	26.2869
652	42 51 04	25.5343	692	47 88 64	26.3059
653	42 64 09	25.5539	693	48 02 49	26.3249
654	42 77 16	25.5734	694	48 16 36	26.3439
655	42 90 25	25.5930	695	48 30 25	26.3629
656	43 03 36	25.6125	696	48 44 16	26.3818
657	43 16 49	25.6320	697	48 58 09	26.4008
658	43 29 64	25.6515	698	48 72 04	26.4197
659	43 42 81	25.6710	699	48 86 01	26.4386
660	43 56 00	25.6905	700	49 00 00	26.4575
661	43 69 21	25.7099	701	49 14 01	26.4764
662	43 82 44	25.7294	702	49 28 04	26.4953
663	43 95 69	25.7488	703	49 42 09	26.5141
664	44 08 96	25.7682	704	49 56 16	26.5330
665	44 22 25	25.7876	705	49 70 25	26.5518
666	44 35 56	25.8070	706	49 84 36	26.5707
667	44 48 89	25.8263	707	49 98 49	26.5895
668	44 62 24	25.8457	708	50 12 64	26.6083
669	44 75 61	25.8650	709	50 26 81	26.6271
670	44 89 00	25.8844	710	50 41 00	26.6458
671	45 02 41	25.9037	711	50 55 21	26.6646
672	45 15 84	25.9230	712	50 69 44	26.6833
673	45 29 29	25.9422	713	50 83 69	26.7021
674	45 42 76	25.9615	714	50 97 96	26.7208
675	45 56 25	25.9808	715	51 12 25	26.7395
676	45 69 76	26.0000	716	51 26 56	26.7582
677	45 83 29	26.0192	717	51 40 89	26.7769
678	45 96 84	26.0384	718	51 55 24	26.7955
679	46 10 41	26.0576	719	51 69 61	26.8142
680	46 24 00	26.0768	720	51 84 00	26.8328

* By permission from *Statistics for students of psychology and education*, by H. Sorenson. Copyright 1936, McGraw-Hill Book Company, Inc.

TABLE U. TABLE OF SQUARES AND SQUARE ROOTS* *(Continued)*

Number	Square	Square root	Number	Square	Square root
721	51 98 41	26.8514	761	57 91 21	27.5862
722	52 12 84	26.8701	762	58 06 44	27.6043
723	52 27 29	26.8887	763	58 21 69	27.6225
724	52 41 76	26.9072	764	58 36 96	27.6405
725	52 56 25	26.9258	765	58 52 25	27.6586
726	52 70 76	26.9444	766	58 67 56	27.6767
727	52 85 29	26.9629	767	58 82 89	27.6948
728	52 99 84	26.9815	768	58 98 24	27.7128
729	53 14 41	27.0000	769	59 13 61	27.7308
730	53 29 00	27.0185	770	59 29 00	27.7489
731	53 43 61	27.0370	771	59 44 41	27.7669
732	53 58 24	27.0555	772	59 59 84	27.7849
733	53 72 89	27.0740	773	59 75 29	27.8029
734	53 87 56	27.0924	774	59 90 76	27.8209
735	54 02 25	27.1109	775	60 06 25	27.8388
736	54 16 96	27.1293	776	60 21 76	27.8568
737	54 31 69	27.1477	777	60 37 29	27.8747
738	54 46 44	27.1662	778	60 52 84	27.8927
739	54 61 27	27.1846	779	60 68 41	27.9106
740	54 76 00	27.2029	780	60 84 00	27.9285
741	54 90 81	27.2213	781	60 99 61	27.9464
742	55 05 64	27.2397	782	61 15 24	27.9643
743	55 20 49	27.2580	783	61 30 89	27.9821
744	55 35 36	27.2764	784	61 46 56	28.0000
745	55 50 25	27.2947	785	61 62 25	28.0179
746	55 65 16	27.3130	786	61 77 96	28.0357
747	55 80 09	27.3313	787	61 93 69	28.0535
748	55 95 04	27.3496	788	62 09 44	28.0713
749	56 10 01	27.3679	789	62 25 21	28.0891
750	56 25 00	27.3861	790	62 41 00	28.1069
751	56 40 01	27.4044	791	62 56 81	28.1247
752	56 55 04	27.4226	792	62 72 64	28.1425
753	56 70 09	27.4408	793	62 88 49	28.1603
754	56 85 16	27.4591	794	63 04 36	28.1780
755	57 00 25	27.4773	795	63 20 25	28.1957
756	57 15 36	27.4955	796	63 36 16	28.2135
757	57 30 49	27.5136	797	63 52 09	28.2312
758	57 45 64	27.5318	798	63 68 04	28.2489
759	57 60 81	27.5500	799	63 84 01	28.2666
760	57 76 00	27.5681	800	64 00 00	28.2843

* By permission from *Statistics for students of psychology and education,* by H. Sorenson. Copyright 1936, McGraw-Hill Book Company, Inc.

TABLE U. TABLE OF SQUARES AND SQUARE ROOTS* (*Continued*)

Number	Square	Square root	Number	Square	Square root
801	64 16 01	28.3019	841	70 72 81	29.0000
802	64 32 04	28.3196	842	70 89 64	29.0172
803	64 48 09	28.3373	843	71 06 49	29.0345
804	64 64 16	28.3549	844	71 23 36	29.0517
805	64 80 25	28.3725	845	71 40 25	29.0689
806	64 96 36	28.3901	846	71 57 16	29.0861
807	65 12 49	28.4077	847	71 74 09	29.1033
808	65 28 64	28.4253	848	71 91 04	29.1204
809	65 44 81	28.4429	849	72 08 01	29.1376
810	65 61 00	28.4605	850	72 25 00	29.1548
811	65 77 21	28.4781	851	72 42 01	29.1719
812	65 93 44	28.4956	852	72 59 04	29.1890
813	66 09 69	28.5132	853	72 76 09	29.2062
814	66 25 96	28.5307	854	72 93 16	29.2233
815	66 42 25	28.5482	855	73 10 25	29.2404
816	66 58 56	28.5657	856	73 27 36	29.2575
817	66 74 89	28.5832	857	73 44 49	29.2746
818	66 91 24	28.6007	858	73 61 64	29.2916
819	67 07 61	28.6082	859	73 78 81	29.3087
820	67 24 00	28.6356	860	73 96 00	29.3258
821	67 40 41	28.6531	861	74 13 21	29.3428
822	67 56 84	28.6705	862	74 30 44	29.3598
823	67 73 29	28.6880	863	74 47 69	29.3769
824	67 89 76	28.7054	864	74 64 96	29.3939
825	68 06 25	28.7228	865	74 82 25	29.4109
826	68 22 76	28.7402	866	74 99 56	29.4279
827	68 39 29	28.7576	867	75 16 89	29.4449
828	68 55 84	28.7750	868	75 34 24	29.4618
829	68 72 41	28.7924	869	75 51 61	29.4788
830	68 89 00	28.8097	870	75 69 00	29.4958
831	69 05 61	28.8271	871	75 86 41	29.5127
832	69 22 24	28.8444	872	76 03 84	29.5296
833	69 38 89	28.8617	873	76 21 29	29.5466
834	69 55 56	28.8791	874	76 38 76	29.5635
835	69 72 25	28.8964	875	76 56 25	29.5804
836	69 88 96	28.9137	876	76 73 76	29.5973
837	70 05 69	28.9310	877	76 91 29	29.6142
838	70 22 44	28.9482	878	77 08 84	29.6311
839	70 39 21	28.9655	879	77 26 41	29.6479
840	70 56 00	28.9828	880	77 44 00	29.6648

* By permission from *Statistics for students of psychology and education*, by H. Sorenson. Copyright 1936, McGraw-Hill Book Company, Inc.

TABLE U. TABLE OF SQUARES AND SQUARE ROOTS* (*Continued*)

Number	Square	Square root	Number	Square	Square root
881	77 61 61	29.6816	921	84 82 41	30.3480
882	77 79 24	29.6985	922	85 00 84	30.3645
883	77 96 89	29.7153	923	85 19 29	30.3809
884	78 14 56	29.7321	924	85 37 76	30.3974
885	78 32 25	29.7489	925	85 56 25	30.4138
886	78 49 96	29.7658	926	85 74 76	30.4302
887	78 67 69	29.7825	927	85 93 29	30.4467
888	78 85 44	29.7993	928	86 11 84	30.4631
889	79 03 21	29.8161	929	86 30 41	30.4795
890	79 21 00	29.8329	930	86 49 00	30.4959
891	79 38 81	29.8496	931	86 67 61	30.5123
892	79 56 64	29.8664	932	86 86 24	30.5287
893	79 74 49	29.8831	933	87 04 89	30.5450
894	79 92 36	29.8998	934	87 23 56	30.5614
895	80 10 25	29.9166	935	87 42 25	30.5778
896	80 28 16	29.9333	936	87 60 96	30.5941
897	80 46 09	29.9500	937	87 79 69	30.6105
898	80 64 04	29.9666	938	87 98 44	30.6268
899	80 82 01	29.9833	939	88 17 21	30.6431
900	81 00 00	30.0000	940	88 36 00	30.6594
901	81 18 01	30.0167	941	88 54 81	30.6757
902	81 36 04	30.0333	942	88 73 64	30.6920
903	81 54 09	30.0500	943	88 92 49	30.7083
904	81 72 16	30.0666	944	89 11 36	30.7246
905	81 90 25	30.0832	945	89 30 25	30.7409
906	82 08 36	30.0998	946	89 49 16	30.7571
907	82 26 49	30.1164	947	89 68 09	30.7734
908	82 44 64	30.1330	948	89 87 04	30.7896
909	82 62 81	30.1496	949	90 06 01	30.8058
910	82 81 00	30.1662	950	90 25 00	30.8221
911	82 99 21	30.1828	951	90 44 01	30.8383
912	83 17 44	30.1993	952	90 63 04	30.8545
913	83 35 69	30.2159	953	90 82 09	30.8707
914	83 53 96	30.2324	954	91 01 16	30.8869
915	83 72 25	30.2490	955	91 20 25	30.9031
916	83 90 56	30.2655	956	91 39 36	30.9192
917	84 08 89	30.2820	957	91 58 49	30.9354
918	84 27 24	30.2985	958	91 77 64	30.9516
919	84 45 61	30.3150	959	91 96 81	30.9677
920	84 64 00	30.3315	960	92 16 00	30.9839

* By permission from *Statistics for students of psychology and education*, by H. Sorenson. Copyright 1936, McGraw-Hill Book Company, Inc.

TABLE U. TABLE OF SQUARES AND SQUARE ROOTS* (*Continued*)

Number	Square	Square root	Number	Square	Square root
961	92 35 21	31.0000	981	96 23 61	31.3209
962	92 54 44	31.0161	982	96 43 24	31.3369
963	92 73 69	31.0322	983	96 62 89	31.3528
964	92 92 96	31.0483	984	96 82 56	31.3688
965	93 12 25	31.0644	985	97 02 25	31.3847
966	93 31 56	31.0805	986	97 21 96	31.4006
967	93 50 89	31.0966	987	97 41 69	31.4166
968	93 70 24	31.1127	988	97 61 44	31.4325
969	93 89 61	31.1288	989	97 81 21	31.4484
970	94 09 00	31.1448	990	98 01 00	31.4643
971	94 28 41	31.1609	991	98 20 81	31.4802
972	94 47 84	31.1769	992	98 40 64	31.4960
973	94 67 29	31.1929	993	98 60 49	31.5119
974	94 86 76	31.2090	994	98 80 36	31.5278
975	95 06 25	31.2250	995	99 00 25	31.5436
976	95 25 76	31.2410	996	99 20 16	31.5595
977	95 45 29	31.2570	997	99 40 09	31.5753
978	95 64 84	31.2730	998	99 60 04	31.5911
979	95 84 41	31.2890	999	99 80 01	31.6070
980	96 04 00	31.3050	1000	100 00 00	31.6228

* By permission from *Statistics for students of psychology and education*, by H. Sorenson.   Copyright 1936, McGraw-Hill Book Company, Inc.

INDEX

303

LEVEL OF MEASURE-MENT	NONPARAMETRIC STATIS		
	One-sample case (Chap. 4)	Two-sample case	
		Related samples (Chap. 5)	Independent sa (Chap. 6)
Nominal	Binomial test, pp. 36–42 χ^2 one-sample test, pp. 42–47	McNemar test for the significance of changes, pp. 63–67	Fisher exact prol bility test, pp 104 χ^2 test for two ir pendent samp pp. 104–111
Ordinal	Kolmogorov-Smirnov one-sample test, pp. 47–52 One-sample runs test, pp. 52–58	Sign test, pp. 68–75 Wilcoxon matched-pairs signed-ranks test,† pp. 75–83	Median test, pp. 111–116 Mann-Whitney L pp. 116–127 Kolmogorov-Smir two-sample te pp. 127–136 Wald-Wolfowitz test, pp. 136– Moses test of ex reactions, pp. 145–152
Interval		Walsh test, pp. 83–87 Randomization test for matched pairs, pp. 88–92	Randomization te two independe samples, pp. 1 156

* Each column lists, cumulatively downward, the tests applicable to the given level of measurement. For example, in the case of k related samples, when ordinal measurement has been achieved both the Friedman two-way analysis of variance and the Cochran Q test are applicable.